T0073858

THE CLOCKS ARE TELLING LIES

THE CLOCKS ARE TELLING LIES

TELLING LIES

Science, Society, and the
Construction of Time

SCOTT ALAN JOHNSTON

McGill-Queen's University Press
Montreal & Kingston • London • Chicago

ISBN 978-0-2280-0843-9 (cloth)
ISBN 978-0-2280-0963-4 (ePDF)
ISBN 978-0-2280-0964-1 (ePUB)

Legal deposit first quarter 2022
Bibliothèque nationale du Québec

Printed in Canada on acid-free paper that is 100% ancient forest free
(100% post-consumer recycled), processed chlorine free

We acknowledge the support of the Canada Council for the Arts.
Nous remercions le Conseil des arts du Canada de son soutien.

Library and Archives Canada Cataloguing in Publication

Title: The clocks are telling lies : science, society, and the construction of time / Scott Alan
Johnston.
Names: Johnston, Scott Alan, author.
Description: Includes bibliographical references and index.
Identifiers: Canadiana (print) 20210305460 | Canadiana (ebook) 2021030569X
  | ISBN 9780228008439 (hardcover) | ISBN 9780228009634 (PDF) | ISBN
  9780228009641 (ePUB)
Subjects: LCSH: Time—Systems and standards—History. | LCSH: Time—Social aspects
  History. | LCSH: Horology—History. | LCSH: Time measurements—History. | LCSH:
  Clocks and watches—History.
Classification: LCC QB223 .J64 2022 | DDC 389/.17—dc23

This book was typeset in 10.5/13 Sabon.

Contents

Tables and Figures

TABLES

FIGURES

Acknowledgments

Writing this book has been a seven-year journey for me, and its story and message feel as relevant now as when I began. I am struck, for example, by how the difficulty of accessing accurate timekeeping in the 1880s turned many reasonable, ordinary people away from modern timekeeping. Without access to it, inequality and resentment were the result. The same appears to be true today, not necessarily with respect to timekeeping, but regarding science itself. It is all too easy for science and academe to appear aloof, for expertise to seem smug, and for pseudo-science to look welcoming in contrast. The pyramidologists of Charles Piazzi Smyth's era have become the climate deniers of our time (and the flat earthers are still at it, all these years later). If we are to overcome these threats to civilization – and make no mistake, the consequences of ignoring climate science represent a very real threat to our future – then the proliferation of misinformation must be combatted, and systemic barriers to education, including income inequality, must be removed. As this book's conclusion suggests, we cannot ignore the value of equitable access to reliable information. These issues must become the rallying cry of today's academics and engaged citizens alike. They already are in some circles. But it is a difficult challenge, and one that must be pursued continuously. I applaud those individuals taking up the challenge, and I have hope that their efforts might someday make a better, kinder world.

Speaking of kindness, I am indebted to many people for their help in writing this book. First and foremost, I owe a debt of gratitude to Stephen Heathorn, whose mentorship and support were invaluable throughout the project. John Weaver and Michael Egan also provided expert advice on early versions of the manuscript, as did David Leeson. Thank you also to Dan Gorman, for encouraging me in my studies, and H.V. Nelles for teaching me the value of narrative as an analytical tool.

Collectively, these six historians, directly or indirectly, helped me realize this was a story about people, not about technology, and the book is better for that realization.

I was fortunate to have the financial support of several institutions and funding bodies, including the Ontario Graduate Scholarship (OGS), McMaster University, and the Wilson Institute for Canadian History, without which I would not have been able to carry out the research necessary for this project.

I am indebted to the staff of several archives, including Mark Hurn at the Cambridge Astronomy Library, Karen Moran at the Royal Observatory of Edinburgh, Kathryn McKee at the St John's College Library (Cambridge), Sian Prosser at the Royal Astronomical Society Archives, John Cable at the Institute of Engineering and Technology Archives, Julie Carrington at the Royal Geographical Society Archives, Chris Atkins at the Maidenhead Library, and Krista McCracken at the Shingwauk Residential Schools Centre. I am likewise thankful to the staff at the Archives of Ontario, the Archives of the Royal Society, Archivo General de la Nacion (Dominican Republic), the Bibliothèque de l'Institut de France, the British Library, the BT Archives, Cambridge University Library, Library and Archives Canada, the Library of Congress, the London Metropolitan Archives, the National Archives of the United Kingdom, and the National Archives of the United States of America,

At McGill-Queen's University Press, Richard Ratzlaff, Kathleen Fraser, and the rest of the team made the publishing experience as smooth as I could have hoped for. Similarly, the recommendations of the anonymous peer reviewers undoubtedly made this book better.

This book originated as a PhD dissertation for the Department of History at McMaster University. In addition, some of the ideas in chapter 1 I previously examined in "Making Time: Transnational Networks and the Establishment of Standard Time in Canada and Beyond, 1867–1905," in *Undiplomatic History: The New Study of Canada and the World*, edited by Asa McKercher and Phillip Van Huizen (Montreal and Kingston: McGill-Queen's University Press, 2019): 56–75.

A delightfully large group of colleagues-turned-friends deserve thanks as well, for their advice on the manuscript, but, more important, for their friendship and support – Chelsea Barranger, Samantha Clarke, Oleksa Drachewych, Curran Egan, Kelsey Hine, Mica Jorgenson, Jacqueline Kirkham, Andrew Kloiber, Shay Sweeney, and Alex Zavarise – I love you all.

Lastly, I would like to thank my family. Thank you to my partner, Megan Johnston, whose love and support kept me sane through the entire project, with diversions and pep talks (not to mention her help translating documents from French). Thank you to my siblings Melissa Sillett and Steven Johnston – may we be companions on many adventures to come. And lastly, thank you to my parents, Gary and Rita Johnston, who encouraged and supported me at every turn. Their love is an inspiration. I dedicate this book to them.

Abbreviations

AMS	American Metrological Society
ASCE	American Society of Civil Engineers
AU	astronomical unit (distance between sun and earth)
BAAS	British Association for the Advancement of Science
CO	Colonial Office (Britain)
FO	Foreign Office (Britain)
IGA	International Geodetic Association General Meeting (Rome, 1883)
IGC	International Geographical Congress (Venice, 1881)
IMC	International Meridian Conference (Washington, DC, 1884)
IPAWM	International Institute for Preserving and Perfecting the Anglo-Saxon Weights and Measures
SAD	Science and Art Department (British)
STC	Standard Time Company

THE CLOCKS ARE TELLING LIES

Introduction

Have you ever looked at a clock and wondered who decided what time it is? We all understand at some level that clock time is an artifice, something concocted for the purposes of convenience. Our clocks and watches tell us the time we have all collectively agreed to use, and our societies run themselves to the tune of their ticking, but our timekeeping system is really just one of approximation and consensus. Even today, with atomic clocks and global-positioning satellites (GPSes) that provide the world with timekeeping precise to within billionths of a second, there is still no one true time. Those atomic clocks are subject to the whims of political agreements about, for example, the length of a second or the arbitrary width of a time zone, and we alter timekeeping to fit national boundaries, or for daylight savings. So it is not physics that determines time, but politics. Physics rejects the idea of a single true time anyway. Guided by Albert Einstein's theories of relativity, modern physicists posit instead that time is relative, subject to change according to velocity and gravitational forces. On a human scale, relativity is an effect so imperceptibly small that it barely registers, but it is significant enough that modern satellite systems have to take into account time dilation in order to work properly. Ultimately, Einstein's insight means humanity cannot look to the universe to find a single all-encompassing timekeeping standard.[1] It's up to us, and so time remains, repurposing an old proverb, "a series of lies agreed upon."[2] There is no "truth" in timekeeping, no perfect, universal time waiting out there to be discovered. It's all made up.

This book tells the story of how we made it up. It asks why do we tell time the way we do? Specifically, how did timekeeping become standardized across the globe? After all, it is a comparatively recent phenomenon. Before the nineteenth century, all time was local time. There was no need

Figure 0.1 | Suffolk, Virginia Rail Collision, 1837, caused by a lack of adherence to the rail schedule.

for a clock in Paris to agree with one in Moscow. On foot or horseback, it was impossible to travel fast enough from one town to the next to care that noon was occurring a few minutes – or even hours – earlier or later. There was no equine jetlag, so to speak.

It was only in the mid-nineteenth century that that began to change. The invention of the railway and the telegraph almost singlehandedly created a newly interconnected world, where suddenly the time differences between cities mattered. Telegraphs required time to be carefully coordinated between sender and receiver, and railways risked loss of life without accurate scheduling (see Figure 0.1). In order to manage this chaos, new norms of timekeeping had to be agreed upon.

These new technologies undeniably provided the impetus for standardizing time. But the invention of the train and telegraph alone is not sufficient to explain why human beings solved the challenges of global timekeeping the way they did. Those solutions were not technologically determined, but rather were socially and politically engineered, and that is far more interesting. This is a story about the growing pains of a newly interconnected world, which reached its climax (where timekeeping is concerned) between about 1875 and 1914.

The preconditions for this revolution in timekeeping emerged in the nineteenth century, which, especially in Europe, might be considered the age of inventory, or the era of taking stock. The centuries-long saga of global exploration was ending, so now Victorians were wholesale engaged in surveying and auditing the world's resources.[3] Such activity could take benign forms, as newly professionalizing fields of science worked to standardize everything from systems of weights and measures, to the taxonomy of butterflies, to time itself. Commercial interests surveyed and mapped land, and organized and categorized crops and exports. But these stock-taking activities also had a dark side, in the form of colonial exploitation. Mapping and surveying could be used to produce knowledge that allowed the metropole to appropriate resources at the expense of Indigenous peoples across the globe.[4] The ability to measure time with precision enabled sailors to better determine their longitude at sea, which in turn facilitated overseas colonization. For good or ill (and it was often ill), the world was being measured, organized, categorized, and standardized, everything put in its place, and timekeeping was no exception.

Of course, it was a messy process. Humanity's ambition to put everything in order outpaced their technological ability to do so.[5] National, professional, and commercial competition, along with class inequality and violent conflicts in colonial spaces, limited these efforts at perfection. There was no shortage of ideas about how best to organize the world, but making everyone agree, by persuasion or coercion, was no easy task.

For timekeeping, this meant that in the mid-nineteenth century asking someone for the time might elicit a complicated answer. The problem was not lack of sources: watches and timepieces were widely available, public clocks adorned city halls and railway stations, and bell signals and calls to prayer of different faiths rang out with religious regularity in many places across the globe. At a pinch, the sun or tides might allow a rough estimate of the time. For urban or rural, rich or poor, nation-state or colony, tools for telling time were ubiquitous.

The problem with time, then, was not a dearth of methods to measure it, but rather a bewildering multitude of often-conflicting and -competing times. Clocks were not synchronized, and even the best-made timepieces could not keep ticking out a perfect rhythm for longer than a few weeks. This meant that one clock varied from the next with striking irregularity. To complicate matters further, the question of whether a clock was considered correct was more often a matter of power, politics, and social dynamics than it was one of insufficient technology. Although clocks might differ by accident, they might also do so deliberately, as various professions, religions, cultures, and nations kept different times (not to mention calendars, each based on unique cultural, religious, and astronomical foundations). Temporal uncertainty was the norm, and people questioned practices that we take for granted in the twenty-first century. Why should a clock have twelve hours on its face? Why should the day begin at midnight? Why does a clock in Boston have to correlate with one in Istanbul or Tokyo? Why should the world's time be counted from an imaginary line running through the Greenwich Observatory in Britain? Why have twenty-four time zones and not ten, or none at all? There are no astronomical, or geographical, or indeed any "natural" imperatives that enforce these things as the only possible way to organize time. These were decisions, originally with uncertain outcomes, made by humans in a particular context. How to measure time was a controversial subject that sparked considerable debate, and had no easy answer.

The centrepiece of these debates was the International Meridian Conference (IMC) held in Washington, DC, in 1884. Here, diplomats, scientists, naval officers, and engineers from nearly thirty countries gathered to discuss the creation of a prime meridian and, by extension, the future of global timekeeping and mapmaking. The conference has gained something of a mythological status as the origin point of modern standard time. Popular history describes it as the moment when the grand schemes of reformers such as Sandford Fleming and William Allen came to fruition to create time zones across the world. But this is something of an oversimplification. Standard time as we know it was not birthed fully formed with pomp and circumstance at Washington in 1884. Indeed, some historians have argued that the conference was a failure altogether for time reformers like Allen and Fleming, because, although it established a prime meridian, it did not actually manage to impose time zones or standard time in any form on its signatories.[6] At best, the IMC was only a stepping stone in a long progression of developments towards modern standard time; a start, not an end, to global changes in time measurement. It took until at least the 1940s for standard time to fully envelop the globe. The IMC's historical significance is further diminished in conventional narratives of modern history because it is overshadowed by the other famous conference of 1884, in Berlin. In the new German capital, the major European powers carved up the African continent, formalizing the "Scramble for Africa." In a way, that conclave set the stage for the twentieth century, foreshadowing the great-power rivalries, colonization and decolonization, and the humanitarian crises that have plagued many of the former colonies since Europe's incursions. Next to Berlin, the IMC seems inconsequential.

Yet the events of the IMC should not be underestimated. If Berlin foreshadowed the twentieth century, Washington was more representative of the nineteenth – it was a product of Victorian debates about the nature of expertise, professionalization, and the desire to understand, order, and organize the world. There is much to learn from studying the Washington Conference. It has a lot to teach the careful observer about the world in which it took place. This is what makes it such a fruitful case study. It is easy to forget that 1884 was its own present. The people involved had no idea what was to come, and made their decisions in an already complex world. For them, the IMC was not just a beginning on the road to something greater, and for some participants at least it was not entirely a failure either. It was an experiment in Victorian modernity, and an expression of a lived present.

This book therefore uses a thoroughly historicist method. It seeks to uncover what the Washington Conference meant to its participants in their own time. When 1884 was the present, how and why did people make decisions? How did they perceive their future? What worried them, excited them, and motivated them? The answers obviously involve political and diplomatic considerations, but there is also the social and cultural world in which the delegates at Washington were immersed. Imperial-power relations acted side by side with the lived experiences of the working class. The economics of time decided when pubs closed, what privileges gave access to accurate time, and who cashed in as the clock was commercialized. Meanwhile, scientists, engineers, diplomats, businessmen, and religious authorities competed for control over international timekeeping. New nations attempted to make their voices heard, and colonies struggled to gain a seat at the table. It is a story of complexity, which perhaps explains why Victorians, like Enlightenment thinkers before them, were so concerned with standardization and organization. Their attempt to standardize time was an effort to simplify a complicated world.

Unfortunately for them, as this book asserts, their attempt to standardize timekeeping worldwide actually made it more complex, rather than simplifying it. There are two broad arguments that lead to that conclusion and that inform this volume. The first is that time is a socially constructed form of knowledge. Scientists, engineers, and diplomats at the IMC did not derive their decisions from infallible principles, nor deduce them logically and objectively from natural laws. They worked them out in a rich, dense context, guided by political, national, and especially professional interests. Accordingly, this study is as much a social and cultural history as it is a history of science or technology. In effect, experts in science and other professions (e.g., astronomers, clockmakers, diplomats, engineers) contested with each other and with the rest of the population for authority over time. Individual personalities and their professional context mattered more than national interest or technological imperatives in the debates over time reform. In later years, from the mid-twentieth century on, time standardization would fall under the purview of the nation-state.[7]

But that was not the case in 1884. It was not some disembodied entity called "Britain" or "the United States," for example, sitting in the grand, often-swelteringly hot room at the IMC in Washington in October 1884, but individuals of various skillsets who happened to be from those countries. The deepest fault lines there existed not between nations, but between professions, resulting even in representatives from the same nation undermining each other's positions. Caught up in the rise of scientific professionalization and the current interests of their respective fields, astronomers, engineers, and naval officers, regardless of nationality, debated heatedly with each other about the best way to organize time. These debates spilled over beyond the diplomatic halls of Washington to the public at large, who quarrelled for years over the best methods of keeping time.

This leads us to the second broad argument of the book: that standard time took the form that it did in large part because of a disagreement about the very nature of temporal knowledge: was it a public good to be shared freely? A commodity to be sold? Or a scientific tool to be entrusted only to those with the professional expertise to put it to good use? Although this debate was rarely acknowledged directly by the historical figures involved, their opinions on these questions underlie nearly all of their actions. Again, this fundamental debate was not confined to the IMC itself. The nature of knowledge about time was discussed everywhere from the poor quarters of British cities to Indigenous communities in Canada to astronomical observatories in the United States and around the globe.

So why did these debates complicate rather than simplify timekeeping? The answer lay in the fact that while some reformers sought to universalize and globalize timekeeping, other experts aimed to restrict and control access to it. Activists like Sandford Fleming and William Allen, for example, wanted a simple, broadly useable standard time that would be accessible to every human being everywhere, replacing local time with a universal system available the world over. But other experts, such as the astronomers at the IMC, envisioned universal time as a specialized tool for astronomical, nautical, and scientific endeavours. To them, universal time was universal only in the geographical sense: that it could be used to determine the precise time anywhere on the globe, if you had the correct tools and education. It was not for use by everyone.

The tension between these two conceptions of global timekeeping was further complicated by the fact that, whereas anyone could use the old system of local time by looking at the sun overhead, standard time, because it would have to be the same everywhere, required some form of expert production and distribution.[8] The means of producing time, in other words, was removed from the ordinary person and placed in the hands of astronomers at the Royal Observatory at Greenwich and other observatories worldwide, who would then distribute that time by wire, and not always for free. This meant that the attempt to standardize time universally across the globe paradoxically ended up making access to "true" time inequitable. The realities of limited technology, combined with astronomers' desire for exclusive access, meant that, while the new universal time gained professional and international legitimacy, its reach remained limited among the general population for decades. A by-product of this unequal access to universal standard time was that older, more accessible forms of time were not erased, but instead continued to exist alongside the new standard time. Local timekeepers now had to decide whether to give up their own authority over time to a higher power (Greenwich) or else to reject Greenwich and compete with it, which many of them did, finding new ways to legitimize the source of their own time instead. Creating a new, authoritative universal time based on Greenwich did not therefore simplify timekeeping but created a hierarchy of times, with the new existing alongside old. Which time was "true" and which was telling lies depended on the observer and their context.

A STORY IN FIVE PARTS

The events of this book take place largely between 1875 and 1914, during which time humanity experienced some of the most rapid and widespread changes to timekeeping in history. Although these changes occurred worldwide, our story focuses often on North America, whose unwieldy geographical width served as the primary impetus for standard time, because coordinating railway schedules across the vast continent was so difficult. A sizable part of the volume also concentrates on Great Britain, which, as the home of the prime meridian and the centre of global timekeeping, offers a stark contrast between the official scientific time produced at Greenwich and the unofficial local times used by much of the general public.

But this is a global story, and the centrepiece is an international gathering, the International Meridian Conference (IMC) of 1884 in Washington, DC, where participants from twenty-five nations debated international timekeeping practices. Other parts of the world, and personages from such places, appear frequently throughout the volume, and the debates happening in Britain and North America were echoed around the globe. This was a worldwide revolution, with diverse participants.

Our story begins, in chapter 1, by tracing the idea of standard time from its inception, through various stages of advocacy and lobbying, to the planning and organization of the IMC. The key figure in these activities was Canadian railway engineer Sandford Fleming. Fleming was the primary instigator of the IMC time debates, but, as a railway-man and engineer, he was an outsider to the scientific community whose proposals ultimately prevailed at the conclave. His occupation made it difficult for him to earn a seat at the very conference that he initiated, and he turned in desperation to sometimes-dubious allies to promote his reforms. His challenges highlight how professional context shaped reformers' participation in the time debates.

Chapter 2 shifts from Fleming to the scientific community. It follows three individuals on the periphery of professional science in Britain, Annie Russell (one of the first women to work at Greenwich Observatory), William Parker Snow (a down-on-his-luck navigator and explorer seeking patronage for his timekeeping ideas), and Charles Piazzi Smyth (a senior but eccentric astronomer whose pseudo-scientific beliefs about the Great Pyramid of Giza in Egypt informed his opinions on timekeeping). The precarious positions of these three individuals put the spotlight on the professional boundaries of Victorian science and place the time debates in the context of the intellectual mood of the period. None of these three attended the IMC, but they and their opinions on timekeeping are representative of the context for the gathering and the resulting time debates. Participants at the IMC could not remove themselves from the cultural baggage delineated in this chapter, such as the tension between amateur and professional, and the place of religion in science. There were rules of etiquette that defined and regulated the types of people who were allowed to weigh in on standardized time. Modern science was becoming an insular pursuit, and this shaped the way scientifically inclined delegates understood universal time.

Chapter 3 picks up the narrative at the IMC itself. It is a deep dive into day-by-day events: who said what to whom, what kind of backroom dealing went on, and delegates' social activities. It confirms what chapters 1 and 2 set up: that professional, rather than national difference was at the heart of the debate, and is the key to understanding the IMC's results. The scientific community obtained exactly what it wanted – a universal time for professionals only. Champions of standard time like Fleming were left in the cold.

Chapter 4 moves away from politics and diplomacy to explore how the time-reform debates were seen by the general public in Britain. It asks who had access to accurate time, and who did not. This is not a comprehensive discussion of all the ways timekeeping affected British society but instead offers a few illustrations of the conflicting ways in which the IMC decisions percolated into the public sphere. The evolution of a time-selling industry, including the careers of Maria and Ruth Belville, women who sold Greenwich time door to door, forms a central part of this chapter, which examines how accurate time was commodified in the wake of the IMC. This chapter is all about legitimacy and the construction of authority, as some of the public mocked universal time as excessive while others coveted it as a symbol of status and modernity.

Chapter 5 examines the dramatic changes in timekeeping in the late-nineteenth-century United States and Canada. It looks closely at reformers' attempts to reshape public norms of behaviour through education. They failed to enforce standard time by law (although they tried, as we see in a court case over a pub's closing time in London, Ontario), so they turned instead to schools to adjust public behaviour. In facilities ranging from prestigious universities to small-town schoolhouses and Indigenous communities, the curriculum and the schedule and structure of the day itself disseminated standard time. The results were mixed, leaving the public with a complicated relationship to standard time well after its introduction in 1883.

The final picture revealed in this story is a world in which timekeeping was more complicated than ever after the IMC, rather than simplified. The conference created a hierarchy of times that existed side by side, and sparked debates over which was the true one. A person's stance on the subject was shaped by their professional context, but also by whether they considered time a public good, a commodity, or a specialized tool. Unequal access to authoritative Greenwich time compounded these divides, leaving the world's timekeeping in limbo. It would stay that way until the invention and implementation of wireless broadcasts in the mid-1920s, which allowed easier access to Greenwich time. The interceding four decades were a confusing but exciting period for timekeeping, in which the future of time itself was anything but certain.

I

Uneasy Beginnings

On a summer day in Ireland, Sandford Fleming missed a train. So begins one of the more prevalent myths about the origins of standard time.[1] According to the tale, a misprinted railway schedule caused Fleming, a prodigiously talented Canadian railway engineer in his late fifties on holiday, to turn up at the station too late. The error cost him an entire day of travel, and he, with little to do but ponder his misfortune, began to consider how to prevent such inconveniences. How could the variety of local times across Ireland, indeed, across the world, be standardized for everyone's convenience? The incident sparked a notion that Fleming hoped would end for good inconsistent timekeeping and its problems. The result of his musings, the story goes, was twenty-four hour-wide time zones, a system now used almost universally across the globe.

Like many apocryphal tales, this story has a kernel of truth. It comes from a reliable source: Fleming himself described the event in one of his earliest pamphlets on time reform.[2] There is no reason to believe it did not happen, although historian Ian Bartky has uncovered some discrepancies in the details of Fleming's account. There is, for example, some misinformation about the relationship between the date of the train event and the publication of Fleming's first paper on time reform.[3] But dating controversies aside, we can be reasonably sure that the event in some form did happen. Fleming missed that train.

So why call it a myth? It is because, like most origin stories, the missed-train story simplifies a much more complicated sequence of events. It does not matter whether Fleming missed a train or not. Standard time as we know it today was not the result of a single ingenious inventor's lightbulb moment. Fleming, and other claimants to the title of inventor of standard time, such as William Allen, Cleveland Abbe, and Charles Dowd, did not act in a vacuum. Standard time resulted from a process involving these people, but not as lone "inventors." They were more like activists – champions of a cause that they could not implement on their own. Certainly, their accomplishments are noteworthy, but the concept of "invention" misrepresents the nature of their contribution and omits the voices of other players with perhaps almost as large a role. The invention narrative is ahistorical, ignoring the context that shaped standard time and its implementation. The personalities mattered, but so did the social, political, cultural, and technological world in which they were embedded.

Squabbles over credit are natural when something new comes into being. For standard time, they were already well under way in 1904 when William Allen wrote a hit piece about Charles Dowd, claiming that he himself invented standard time and that Dowd's contributions were minimal.[4] Even earlier, there is evidence that Fleming rushed to print his earliest paper on standard time in 1878, to prevent Cleveland Abbe and the American Metrological Society from beating him to the punch.[5] But even if we could pick a "winner," that tells us nothing useful. Dowd undoubtedly came up with the idea first, but was ignored. Fleming's was the one that was ultimately adopted – with modifications – after he joined forces with Cleveland Abbe, who had developed a similar plan. Meanwhile, William Allen did the most to implement that idea on North American railways. European innovators similarly should not be forgotten, like Britain's astronomer royal, George Airy, who helped set up Greenwich time for use by his country's railways in the 1840s and '50s, and Russian astronomer Otto Struve, who pioneered work on the prime meridian in the early 1870s.[6]

The point is, singular moments of invention don't matter. The long, drawn-out process by which the idea of standard time reached the international community is far more interesting. There was nothing inevitable or preordained about the process, and there was considerable opposition. This chapter uses Fleming as our guide into the subject to show how his idea quickly grew beyond him. First, we look at Fleming's early publications and his (often-fruitless) struggle to have scientific authorities notice them. We then examine his widening search for allies on both sides of the Atlantic, where once again he faced resistance from the scientific community and turned instead to fringe organizations and disreputable intellectuals. Finally, we demonstrate how Fleming was nearly excluded from the political process of establishing a common prime meridian for timekeeping. This antipathy was a result of international norms, imperial realities, and interdepartmental rivalries, each part of a ramshackle system of global governance of time that shaped what an independent actor like Fleming could achieve.

ON THE OUTSIDE LOOKING IN

For professional science, Sandford Fleming (1827–1915) (Figure 1.1) was something of an outsider. The late nineteenth century boasted a growing international network of professional scientists. Of these, the astronomical community was most directly responsible for shaping the discussions about global timekeeping. But Fleming was not one of them. He and several of the most fervent proponents of standard time were not particularly welcomed by this scientific network. As an affluent gentleman of Scottish heritage, with useful political connections in Canada and a background in engineering, Fleming was not without economic clout or professional standing. But during the key period of his advocacy for standard time, his career was in flux, as he had been unceremoniously fired from his post as chief engineer for the Canadian Pacific Railway in 1879. Although he did soon find a role as chancellor at Queen's University in Kingston, Ontario, he quickly found out that a background in railway engineering did not grant an easy path into the sciences. Railways, by the late nineteenth century, were tools of business and politics, not academic inquiry, and professional astronomers paid him little mind. These were the very people to whom he needed to appeal if his time-reform ideas were to succeed, but they put him and his proposals through a harsh process of scrutiny, rejection, and, occasionally, ridicule. As a result, his campaign to reform timekeeping was an uphill battle from the start.

Fleming's first rejection came on his initial attempt to share his proposal with a scientific audience. In 1876, shortly after the missed-train incident, he sailed from Ireland to Scotland, the land of his birth, where he attended the annual meeting of the British Association for the Advancement of Science (BAAS) in Glasgow. He planned to give a presentation on uniform time, but, for unclear reasons, he was never given the opportunity.[7] He tried again at the BAAS meeting in Dublin in August 1878, but again failed. Given the pre-eminence of the other speakers at the BAAS conclaves, all paragons of the British scientific community, it is unsurprising that this relatively obscure Canadian didn't make the cut.[8]

Ignored in Britain, he instead printed a version of his paper in French and tried to arrange for an American correspondent of his, Frederick Barnard, president of Columbia College (now University) in New York, to read it at Paris's Exposition Universelle (May–November 1878). Unfortunately, Barnard proved unable to attend.[9] Lightbulb in his head or no, Fleming could not get anyone else to see the light. Defeated, he returned to Canada in the autumn of 1878.

Figure 1.1 | Sandford Fleming, 1895.

Back in North America, Fleming's bad luck began to turn. Through his correspondence with Frederick Barnard, he learned that the American Metrological Society (AMS), of which Barnard was president, was also discussing time reform. It had been doing so since 1873, under the direction of Cleveland Abbe, whose personal interest stemmed from his 1874 attempt to determine the height of the Aurora Borealis above the earth.[10] Volunteers from across the United States took measurements for him, but Abbe could not collate the data, because each observer used a different local time and failed to indicate the error of their timepieces.[11] Abbe began looking for a solution, and the AMS meetings became his outlet for discussing how to standardize time for the purposes of scientific observation.

Fleming was pleased to find someone else taking time reform seriously. For him, the AMS discussions were a promising sign, but he also saw them as a threat. He had self-published his pamphlets on time reform and achieved little circulation, and he worried that the AMS might publish others' papers on the topic before his own work gained recognition. He needed a publication by a respectable scientific society immediately.

Fleming turned to the Canadian Institute, a professional society that he had helped found in Toronto in 1849 as a forum for engineers and surveyors, which later worked to advance science more broadly. Though not involved with it for at least a decade, Fleming found that his prior connection earned him a respectable publication through this professional society. Fleming now became fast allies with Abbe and the AMS, who might otherwise have been his competitors.[12]

Fleming's early timekeeping papers were full of mistakes and overly complicated methods, and his ideas evolved over time, but his core arguments remained consistent. He advocated for three key changes. First, he wanted to replace individual local times with twenty-four standardized zones around the world, each fifteen degrees wide, representing one temporal hour. Second, he proposed a single prime meridian for the world, from which to measure the zones. Meridians were vital to navigation at sea, and there were dozens of them in use in the 1870s, but a single, shared prime meridian had never been agreed on. Fleming did not particularly care where it was placed, although he favoured one hundred and eighty degrees from Greenwich, in the middle of the Bering Strait, because it was a neutral location, passing through no major landmass. Third and finally, he advocated for the use of a twenty-four-hour clock, eliminating a.m. and p.m. (a misprint of a.m. v. p.m. had caused his missed train, and he liked the symmetry of having twenty-four hours in the day to match the twenty-four time zones around the globe).

What Fleming lacked in academic standing, he made up for with political clout, at least in Canada. Through the Canadian Institute, he got his paper forwarded to the governor general of Canada, the Marquess of Lorne, Queen Victoria's son-in-law, who in turn forwarded it to the colonial secretary in Britain and asked that it be distributed to the principal scientific societies of Britain and other nations for comment.[13]

The responses were not particularly encouraging. The colonial secretary ruled out any official action, writing in October 1879: "It has been the custom of Her Majesty's Government to abstain from interfering with recognized usages in questions of social importance until the spontaneous use of any novel system ... has become so extensive as to make it desirable that authoritative regulations should be sanctioned ... and it does not appear that such a condition of affairs ... has yet arisen."[14] In other words, the British government refused to force the public to change how they kept time. It was a decision with serious implications for political philosophy: should a government be able to regulate social norms to such a degree as to rearrange how its subjects measured time itself? In 1879, the answer was decidedly no.[15]

As before, the scientific community was equally sceptical. Members of Britain's Royal Astronomical Society, for example, read the paper at their council meeting, but declined to offer any feedback.[16] The Royal Geographical Society was more amenable, but doubted the feasibility of such a scheme. "There is nothing to be said against the proposal," wrote one member, "except its impracticability, which is such that no scientific body is likely to urge it seriously."[17] The Royal Society of London stated that, although it was disposed to support the idea, "no scheme of the kind would have much chance of success unless there were a general readiness on the part of civilized nations to seriously entertain the question."[18] In other words, there was a near-consensus in 1879 that Fleming's plan for a universal standard time was a pipe dream; laudable, but unachievable.[19]

The astronomer royal, George Airy, and the astronomer royal for Scotland, Charles Piazzi Smyth, were also asked for their opinions. As the highest astronomical authorities in Britain, their opinions carried great weight (Airy's more so than Smyth's, as we see below). Neither was particularly impressed. Smyth predicted that local time would never be replaced, "no matter what beautifully-written schemes any few very learned men may propose in their closets."[20] He condemned Fleming's choice of the Bering Strait for the prime meridian; a fervent British nationalist, he was offended by the thought of placing that line so far from the centre of British civilization: "It is in a part of the world where there are either no inhabitants at all, or, if a few do reside near one end of the line, they are a miserable driblet of wretched Kamchatkan savages, prowling with difficulty for food over snowy wildernesses under the doubtful rule of Russia!"[21] Smyth's racially charged rhetoric was accompanied by equally outrageous political accusations. He called Fleming's global scheme, and internationalism in all its forms, communist nonsense. Fleming was, according to Smyth, "running full tilt against common sense."[22]

Smyth proposed an alternative location for the prime meridian: the Great Pyramid of Giza, in Egypt.[23] The pyramid lay roughly between Britain and its most precious imperial possession, India, but Smyth also liked the spot because he believed that the pyramid held prophetic secrets foretelling the future of the British nation under God (more on this below). Fleming's standard-time proposal contradicted Smyth's somewhat unorthodox conception of the world, and Smyth was loath to support it.

Airy was similarly scathing, though more level-headed: "I set not the slightest value on the remarks extending through the early parts of Mr. Fleming's paper [eliminating local time and replacing it with zone time, using a complicated system of notation that replaced the numbers 1–24 with letters]. Secondly, as to the need of a Prime Meridian, no practical man ever wants such a thing."[24] Airy was not wrong about the last: a prime meridian had never been necessary. A meridian is simply any line of longitude that runs from the north pole to the south pole. For navigation at sea, sailors could pick any meridian to be their baseline. Navigators would set a clock to match the time at their chosen meridian. Then, on their journey, by comparing the time at that meridian to their local time, they could calculate their longitude, and hence their position at sea. It did not matter where on earth their chosen meridian was located, so charts and almanacs picked specific meridians for convenience. Observatories made the best meridians, because they had the best equipment to accurately confirm the local time, but navigators could simply use the last point of land they saw, or their last port of departure.[25] No one had ever seriously considered a worldwide time-keeping system, which would require a prime meridian to avoid multiple competing standard times (one for each meridian in use).

Airy had previously seen one of Fleming's early unpublished papers on time reform, back in February 1878.[26] Less than enthused, he wrote privately that public habits, not government, must effect this change.[27] He pointed out that the British public had already adjusted its habits once, after the railways adopted Greenwich time in the 1840s and 1850s, not enforced by law, but implemented merely for convenience (Greenwich time was not made legal time in Britain until thirty years later, in 1880).[28] Airy had been instrumental in setting up the time signals that permitted British railways and telegraphs to adopt Greenwich time, but he was not interested in extending its pre-eminence internationally, and certainly not via government coercion.[29]

FLEMING LOOKS FOR ALLIES

Rebuffed again by scientific societies in western Europe, Fleming changed tactics. In one way, he took Airy's advice, and looked to the railways. It was in the transportation business, not in government and science, that his ideas for time reform would find their first home.

The years 1880–83 saw an incredible flurry of activity for Fleming: letter writing, networking, petitioning, preparing and administering surveys, and attending conferences. In this barrage, Fleming was one voice among many. Cleveland Abbe, William Allen, Frederick Barnard, and Thomas Egleston (an American engineer) all played major roles. This campaign suggests strongly that standard time was a product not of invention, but of promotion. Advocacy and activism pushed the topic from the obscurity of utopian scheming into the realm of possibility.

The forum for these campaigns and discussions was professional societies. Fleming's Canadian Institute was involved jointly with Abbe and Barnard's American Metrological Society, but Fleming joined several other groups as well, presenting his ideas to the American Society of Civil Engineers (ASCE) and the American Association for the Advancement of Science (AAAS).[30] Of course, only a handful of members in any of these societies ever paid time reform much attention.[31] But these individuals' voices were amplified by the tacit support of a professional society.

Fleming convinced the ASCE to establish a standard-time committee, with him as its chair. He immediately set to work, using the society's distribution network to circulate questionnaires to engineers and railway managers in Mexico, Canada, and the United States, as well as to interested academics.[32] Respondents overwhelmingly agreed that railways ought to reshape their timekeeping methods but disagreed about how. Many, for example, were unconvinced about changing local times to align with railway time. But the survey results made it clear that North America's railway network badly needed a more manageable time system.[33] Armed with these findings, the ASCE went on to petition the U.S. Congress to call an international conference to establish a global time system. Acting in tandem with the ASCE, the AMS also began lobbying Congress. At first, Abbe felt it was unlikely that the American government would call such a meeting and hoped that Canada might do it instead.[34] However, after the governor general's failure to entice any action on the part of the British, the AMS looked inward.

The AMS was undoubtedly the most respectable and politically influential of the U.S. societies pushing for time reform. As Egleston explained to Fleming in early 1883, the ASCE alone could not achieve much, as engineers had little clout. By joining with the AMS, however, it could "secure the interest and cooperation of the most powerful men in the country, which might not be so certain if we worked with the civil engineers alone."[35] The lobbying process was complicated, and internal conflicts between the U.S. Signal Service and the U.S. Naval Observatory heavily affected the outcome.[36] Meanwhile, not all of the time reformers agreed with the AMS's ideals. The time-reform movement had breadth, but that also created plenty of room for fractures and infighting.

Through the AMS, Fleming eventually reached "the most powerful men in the country," but it took time. In seeking allies, he cast his net wide following his failures in Britain and France, and he eagerly courted any society or organization that showed an interest. The AMS was his luckiest catch, but he also became entangled with some less reputable organizations.

The best example of this was the International Institute for Preserving and Perfecting the Anglo-Saxon Weights and Measures (IPAWM).[37] Formed in 1879 by a railway engineer named Charles Latimer, this body, based in Cleveland, Ohio, ostensibly advocated for continuing use of British (imperial) weights and measures. Such societies were not uncommon, given the era's predilection to quantify, categorize, and measure anything and everything.[38] To measure something was to know it: classifying species, calculating the rotation of astronomical bodies around the sun, or surveying land claims and longitudes was creating "knowledge" of a particular kind. Such careful measuring and mapping were not merely aloof scientific endeavours, either. They had real-world ramifications, and often enabled colonialism.[39] The IPAWM was an outspoken, though marginal, part of this global movement of quantification.

Calling it a "movement" should not imply cohesion: divisions and debates about how best to measure the physical world were the norm. The IPAWM was particularly antagonistic towards the French metric system, which was rapidly gaining global acceptance as a "perfect," supposedly impartial system. The IPAWM, however, saw it as anathema to the perfection of measurement, believing it to be based on poorly calculated standards, while the British system was ordained by God.

The IPAWM's support for the British system was not unusual, but its line of reasoning was unorthodox. Its president, Charles Latimer, was a fervent disciple of Charles Piazzi Smyth's pyramidology. According to these theories, Israelite slaves had built the Great Pyramid of Giza with divine inspiration from God. Smyth, and other adherents, claimed to have found evidence in the measurement of the pyramid that the British inch was used in its construction, and therefore British measures must be divinely inspired.

Smyth's membership explains the "international" in the IPAWM's name, but most members were Americans, and its activities centred around Charles Latimer in Cleveland.[40] Nonetheless, the body caught Fleming's attention in 1881 when Latimer, who also belonged to the ASCE, presented a paper on the pyramid and weights and measures. Both Egleston and Barnard asked Fleming to shut down its absurd arguments.[41] Barnard called the IPAWM a "reactionary society" and shared with Fleming his own papers that refuted Smyth's and Latimer's claims concerning the Great Pyramid.[42] Egleston wrote to Fleming that the IPAWM was "an organization without standing whose action has been particularly discreditable, and who have mentioned Dr. Barnard in communications to the United States government most disrespectfully by name. I happen personally to know the moving spirits of this society and they are men with whom affiliation is almost impossible."[43] In fact, Barnard and Abbe disliked the IPAWM so much that, when Fleming suggested including it in joint actions with the AMS and the ASCE, these societies delayed their plans to lobby for an international conference on time reform while their committees tried to avoid such collaboration.[44] Barnard, Egleston, and Latimer may have agreed on many aspects of time reform, but hostilities between them ran too deep for cooperation. Barnard in particular was a fervent supporter of the metric system, and had little patience for Latimer's wild accusations.

However, Fleming kept in contact with the IPAWM, because Latimer showed an interest in standard time, and Fleming needed allies. In late 1881, Latimer invited Fleming to join the IPAWM's new standard-time committee. About the same time, Latimer wrote to Smyth about him, saying that although Fleming worked with Barnard, there was no indication that he agreed with him on the metric system. Latimer also told Smyth (correctly) that Fleming was not necessarily attached to the Bering Strait as the prime meridian, a position that Smyth had denounced a few years earlier.[45] Then, in December 1882, Latimer unilaterally enrolled Fleming in the IPAWM, adding his hope that the new affiliate might become anti-metric.[46]

Fleming, though never committing to Latimer's more fanciful ideals, continued to be active in IPAWM affairs, serving on the standard-time committee, answering a questionnaire on time reform, and contributing articles to the body's publications.[47] When U.S. (and Canadian) railways agreed to begin using time zones at noon on 18 November 1883, Latimer congratulated Fleming, complaining that Dowd and Allen were receiving all the credit, when really it belonged to Fleming.[48]

Fleming's involvement with the IPAWM was ultimately fruitless. In fact, it actually slowed his progress with the ASCE and the AMS. But after such early rejection, he joined any organization that would support him, even a disreputable one, especially one so willing to sing his praises as the IPAWM. Elsewhere, however, time reform continued to meet with roadblocks.

After his initial failures in Britain and France, Fleming had focused on campaigning in North America. But Barnard still had contacts overseas and tried to secure "some declaration in favor of our scheme from the International Association for the Reform and Codification of the Laws of Nations," he told Fleming in August 1881.[49] The response was lukewarm at best. A more promising opportunity came in September 1881, when the International Geographical Congress (IGC) met in Venice. This professional gathering of geographers had met twice before in the 1870s and discussed the idea of a prime meridian for navigational purposes, but had ultimately shelved it for future study.[50] Barnard and Fleming hoped to bring that topic back into the limelight.

In preparation for Venice, Barnard attempted one last time to convince Astronomers Royal Smyth and Airy to see reason. Smyth told him that his opinion had not changed.[51] Airy was just as stubborn: "What does a man living in Ireland or Turkey care about Cosmopolitan Time: it is wanted by sailors, whose profession carries them through great ranges of longitude … and there its utility ends."[52] Frustrated, Barnard complained that Airy had not properly understood Fleming's paper (Airy had said as much in his letter) and that he had "little hope of bringing such a man around by talk after he had well committed himself. He is no doubt a great man, but he is excessively opinionated, and he is sometimes mistaken, as he was most lamentably in the case of Adams and the planet Neptune."[53] In the 1840s, Airy had infamously claimed that that Cambridge astronomer J.C. Adams (later a delegate to the International Meridian Conference [IMC] in 1884) had discovered the planet Neptune, when in fact a French mathematician, Urbain le Verrier, had done so, after Adams had predicted its existence and position. Barnard, upset by the obstinacy of what he saw as a flawed old man, told Fleming that "both of them [Smyth and Airy] were disappointing to me."[54]

Unable to sway Britain's leading astronomers, Fleming went to Venice alone in September 1881. Abbe and Barnard sent letters to be read by their representatives. The IGC received Fleming's proposals with mild interest, but not enough to make much impact.[55] He did convince Italian geographers to lobby their government to hold another international congress, but very weak resolutions followed. The topic stayed in subcommittee: no vote was taken by the whole conference. Both indifference and outright hostility derailed the reformers' efforts.[56] The Italian government was slow to respond and by March 1882 had done nothing.[57] Fleming at this point turned his attention towards the planned Washington Conference on an International Prime Meridian, for which the AMS had successfully lobbied, at long last.[58]

In October 1883, Italy hosted the International Geodetic Association's (IGA) General Meeting in Rome. The prime meridian was once again on the table for discussion. A disillusioned Fleming did not attend, writing to his friend Charles Tupper, then Canada's minister of railways and canals, "from what I know of similar meetings which have over and over been held in the cities of Europe I do not anticipate any satisfactory results or any results at all beyond postponing a settlement of the question indefinitely. The people of this country [Canada] are more practical."[59] But in Rome Fleming would have gained an unexpected ally. The British delegation included William Christie, the new astronomer royal who had just replaced Airy. Christie, for his own reasons, was far more amenable to the idea of a prime meridian and would campaign for it behind the scenes at the IGA and beyond.

With Christie's influence, the IGA conference in Rome passed several promising resolutions, including the selection of an international prime meridian at Greenwich and formal approval of the U.S. proposal to hold a diplomatic conference in Washington in 1884, to ratify this new common meridian.[60] But the Rome Conference was a gathering of scientists, not diplomatic representatives, so could not make binding arrangements. Moreover, it had complicated the issue by suggesting a controversial trade-off: if other countries abandoned their own meridians in favour of Greenwich, Britain in return ought to move in favour of the metric system, or at least pay its share of the costs to the Metre Convention of 1875, which had set up a bureau to standardize and verify comparisons between different systems of measure. The British delegation, led by Christie, did not reject the metric proposal outright, and was even open to Britain's contributing to the Metre Convention.[61]

Regrettably for Fleming, Christie lacked his government's support, and there was political backlash over paying anything to the Metre Convention.[62] Moreover, while Fleming wanted universal time zones for everyone, the Rome Conference laid out a specialized time system for railways, ships, telegraphs, and observatories.[63] Its recommendations meant that universal time was to be a specialized tool for the scientific community, not a new public norm for use in ordinary life.

Back in North America, the time reformers were having better luck. An aggrieved Fleming had suggested somewhat unfairly that people from his continent were more "practical" than Europeans. But the notion of these forward-thinking North Americans being held back by Old World conservatism is nonsense.[64] There were some (like astronomer Simon Newcomb) who opposed time reform, just as there were Europeans who supported it.[65] The Russian astronomer Otto Struve, for example, championed the idea of a prime meridian (for navigation) long before Fleming, and one of Fleming's staunchest supporters was a Spanish naval officer, Juan Pastorin (who would attend the IMC). Perhaps some imperial arrogance on the part of European scientists, and British academics in particular, meant they downplayed the ideas of a colonial subject like Fleming. But scholars from the colonies and dominions did exist, and some rose to prominence, so long as they conformed to European conceptions of scholarship.[66] Imperial prejudice alone does not explain Fleming's failures in Europe, which were also the result of professional differences. Fleming was an engineer, not a scientist; even in North America his reach was greatest with engineers and railway officials, not scientists.

Enter William Allen. A latecomer to time reform, Allen was an American railway engineer and editor of the *Travelers' Official Railway Guide for the United States and Canada*. He learned in late 1881 about the AMS and its joint efforts with the Canadian Institute and the ASCE and jumped on board. While the AMS continued to petition Washington for an international conference on time reform, Allen looked to the railways to make more immediate changes. It was he who, as secretary of the General Time Convention, brought about the use of time zones on all railways in the United States and Canada. The Convention's twice-yearly meetings originally coordinated schedules between railway companies, but Allen directed their attention towards a shared system of time zones instead. In October 1883, while the Rome Conference was under way, Canadian and U.S. railways, meeting in railway hub Chicago, adopted standard time, using Greenwich as a prime meridian, to begin at noon on Sunday 18 November 1883.

This event has been the subject of myth as well. Each year in the 1940s and '50s the Association of American Railroads (successor to the General Time Convention) grandly celebrated its anniversary as the "day of two noons," the date when clocks across the continent were reset and standard time was born.[67] But the change on 18 November 1883 was not so momentous as all that. It was not truly the birth of universal standard time, as it did not apply to the whole globe, nor did it even apply to all aspects of life in the two countries. It affected only railway time, not local time. In many places, the local clocks were not changed, and travellers now had to set their watches upon arrival at the railway station. Numerous major cities did adjust their local time to match railway time, but not all.

In cities that did change their clocks, critics were quick to object.[68] One complainant from Indianapolis, for example, bemoaned that "the Sun is no longer to boss the job. People – 55,000,000 of them [i.e., Americans] – must eat, sleep and work as well as travel by railroad time. It is a revolt, a rebellion. The sun will be requested to rise and set by railroad time. The planets must, in the future, make their circuits by such timetables as railroad magnates arrange. People will have to marry by railroad time, and die by railroad time. Ministers will be required to preach by railroad time ... We presume the sun, moon and stars will make an attempt to ignore the orders of the Railroad Convention, but they, too, will have to give in at last."[69] Other critics – experts and laypersons alike – offered counterproposals. An anonymous Toronto woman, for example, wrote to Fleming on 19 November 1883, the day after the railway adoption, with her own scheme. She suggested a twelve-hour day, with each hour doubled in length to 120 minutes. In this way, as she put it "then midday would be the sixth hour as in the olden times, among the Jews; and the story of the most momentous event which has ever transpired in all time would read truly again."[70] This woman's faith-driven proposal demonstrates that for some people the change was deeply personal – an existential question; more than just an inconvenient watch adjustment.

Fleming, for his part, did not even note the "day of two noons" in his diary. For him, the work was not yet over. He had larger goals. The first was to make the new railway-zone time into legal time in the United States and Canada, eliminating local time forever. The second was to extend zone time worldwide. He would find both tasks extraordinarily challenging.

MAKING THE TIME CONFERENCE A REALITY

Fleming's attempts to make standard time into legal time did not go particularly well. His most useful precedent was the act passed in 1880 in Great Britain, making Greenwich time the legal time in England, Scotland, and Wales, and Dublin time the legal time throughout Ireland.[71] But Greenwich time was not zone time, as Fleming's critics pointed out. In the United Kingdom, the Irish Sea provided a natural barrier between the two time standards, avoiding an abrupt time change. In North America, such convenient natural geographical divisions were rare. In the United States, the question also became entangled with states' rights and residual powers: should the federal government declare legal time, or the states? In 1882 Connecticut had taken it upon itself to declare that railway time was legal time. Fleming considered it a victory, but worried that it set a difficult precedent. Federal action was a surer path to a universal system across the United States. Piecemeal state-by-state legislation would be a long slog. But the national government was slow to act: as late as 1892, it had established standard time as legal time only in the District of Columbia.[72]

Fleming's attempt to extend standard time overseas moved equally slowly, but by August 1882 the AMS had at last convinced the U.S. government to convene a conference on the subject. A circular was forwarded to all countries with diplomatic relations with the United States, asking whether such a gathering would be met with approval. When positive answers were received (it took some time; the British government was particularly slow in responding), the secretary of state sent out invitations in December 1883. The conclave – eventually called the International Meridian Conference (IMC) – was to meet at Washington, DC, the following autumn, in October 1884.

As one of the chief activists in the time-reform movement, Fleming seemed an obvious choice to represent Canada. But just as he had been rejected by the scientific community, he was nearly excluded from the political sphere as well. The vagaries of international politics, Canada's colonial/dominion status, interdepartmental rivalries, and a lack of communication almost cost him his seat at the table. Despite Fleming's modern reputation as the father of standard time, he was, at the time, disposable. Seen from Europe, he was simply a wealthy railwayman from the colonies who harboured utopian ideas. His application to join the international conference led to an uphill battle.

Fleming began lobbying early, almost a year before the conference was called. In early 1883, he asked both the Canadian Institute and the Royal Society of Canada to send memorials to the governor general, Lord Lorne, requesting that Canada be represented. The viceroy's reply was guarded. Canada, being a dominion under the crown, did not have direct diplomatic relations with the United States; its participation depended on the goodwill of Britain's Foreign Office. But Lorne did promise (on 8 May 1883) that if Canada secured an invitation, either directly or via the United Kingdom, a representative would be sent.[73]

Not content with that answer, Fleming wrote the next day to Charles Tupper in London, where he was the Canadian high commissioner (quasi-ambassador) in London. "It does not appear," Fleming wrote him, "that Canada has been invited to take part in the international conference proposed to be held at Washington, but no country in the world is more interested in a satisfactory solution to the problem than Canada and as a matter of fact the movement for the solution originated in Canada."[74] Fleming was of course referring to his own early publications with the Canadian Institute. He asked Tupper to try and secure Canada representation.

As a dominion, Canada could not be invited unless by a specific act of the U.S. Congress, and also by the agreement of Her Majesty's (British)

Government. A more likely solution was to have Canada be represented as a part of the British delegation. On 9 June 1883, after a campaign of letter-writing, Fleming was told that, were Canada to be invited, he would be its delegate.[75] He was assured his place, but only if Canada itself had a seat at the table.

In London, Tupper set to work on Fleming's behalf. The process was messy, in large part because it was unclear which government department had jurisdiction. The Colonial Office (CO) was involved, as it was in all of Canada's relations with Britain, but a diplomatic conference was also a matter of foreign policy, which meant the Foreign Office (FO) had a hand in the choice of delegates. But the Science and Art Department (SAD), a branch of the Board of Trade, was involved in all things to do with science and technology. When British scientists had been sent as delegates to the Rome Conference in the autumn of 1883, the SAD had selected them. As the Washington Conference was to cover similar topics, it made sense that the SAD would take part in the decision-making process again. These three departments, each with its own agenda, existed in parallel without a clear hierarchy. And there was one final layer of bureaucratic complexity – money. None of these departments could act without the oversight of the Treasury, which had to approve funding for the delegates' travel costs.

When the CO received word in spring 1883, via the governor general and Tupper, that Canada wished to be represented, it forwarded the request on to the FO. But the FO hesitated. There were two problems. First, the United States had not yet issued official invitations; in the autumn of 1882 it had sent merely an inquiry as to whether Britain might be interested in attending if such a conference were to be held. The inquiry, received in November 1882, did not even include a date for the proposed event. Second, the Treasury had already decided internally not to fund delegates for such a conference.[76] As a result, the FO offered an alternative. Its officials had heard that the SAD was involved with organizing the Rome Conference, and asked whether Fleming might like to attend Rome instead.[77] On 9 June 1883 the CO replied in the negative, stating that Fleming wished to attend Washington, not Rome, and requested that the FO ask the British ambassador in Washington, Lionel Sackville-West, to take steps to ensure Fleming's recognition as a delegate. West told U.S. secretary of state Frederick Frelinghuysen about Fleming's request for representation. Frelinghuysen replied that he would be pleased to recognize Fleming, but only after Great Britain confirmed its intention to participate.[78] In other words, Canada would not be recognized individually; Fleming was to be a British delegate, or nothing. Given the Treasury's decision the previous November, it seemed likely that no representative of the British Empire would attend at all.[79]

In early July 1883, having heard that the United Kingdom might not attend Washington, Fleming forwarded via Tupper a letter urging it to participate. He stressed that Canada and the United States were willing to accept Greenwich as a prime meridian, an outcome that just might be inevitable if only the British would attend. Fleming suggested that only with their participation could this "problem which has long embarrassed geographers, astronomers and navigators" be settled.[80]

Fleming's pleas were greeted with indifference over the summer of 1883. Without further details from the United States, Lord Granville at the FO was content to leave the matter to rest, awaiting the results of Rome before considering Washington any further.[81] Once the Rome Conference was over (it took place October 1883), Fleming did not waste the chance to bring up the subject again. He hoped that the promising outcome of Rome, as well as the adoption of railway standard time in the United States and Canada, might entice more interest. To broach the subject, he turned once again to Charles Tupper. The two men shared both a friendship and a working relationship, having known each other since at least 1864, when Tupper, as premier of Nova Scotia, had appointed Fleming chief engineer of the Nova Scotia Railway. This ultimately led to Fleming being named chief engineer of the Canadian Pacific Railway too, although Tupper (as minister of railways and canals) was later forced to remove Fleming in 1879 because the project went over budget. Still, the men remained close friends, and bonded over their sons both volunteering to fight against the North West Rebellion in 1885.[82] The extent of Tupper's admiration for Fleming is evident in the minutes accompanying Fleming's letters that autumn, where Tupper scribbled a note suggesting Fleming deserved a knighthood, an honour that he received in 1897.[83]

In early November 1883, Tupper forwarded Fleming's letters on to the CO. In them, both men emphasized that Greenwich might become the world's prime meridian. They pointed out that the recent events on North American railways made the United States an ally in the fight for Greenwich, and that Russia was on board as well. "I anticipate but one result at Washington," said Fleming, "if Her Majesty's Government will only accept the invitation to participate."[84] Despite his best efforts, however, the matter was not taken up until the New Year.

By then, the United States had finally sent out proper invitations, with the date set for 1 October 1884. In January 1884, Britain's Science and Art Department sent off its report on the Rome Conference, and was waiting on responses before forming a committee to discuss a Washington delegation.[85] The FO, with the SAD's report on Rome and the official U.S. invitation in hand, had to make a final decision on Britain's attendance. It consulted the Treasury, but it, having already said no once, was not very willing. It asked Granville whether there were any "political grounds" for accepting the invitation.[86] The FO admitted that the Washington Conference had little political significance, but that the Treasury ought to ask the CO and the SAD. Both considered the subject (the prime meridian) important.[87] The CO, on behalf of Fleming, replied that, yes indeed, delegates ought to be sent.[88]

Then the Treasury, for an unknown reason, asked the Royal Society, instead of the SAD, for its opinion. This confused and annoyed SAD secretary John Donnelly: "What a way of doing business. They never told us – I shall see if my lads won't rub their noses in it," he wrote.[89] Despite the snub, Donnelly was at least pleased that the Royal Society had not sunk their chances. The august body, despite having rejected Fleming's papers on time reform back in 1879, now wrote to the Treasury that the matter was worth pursuing. Its reply, drafted at a Council meeting, said: "There appears a very good prospect that the meridian of Greenwich will be adopted by the civilized world if our government be duly represented at the Washington Conference. But if the nation which far more than any other is interested in the selection of a prime meridian should not think it worth while to take any steps in the matter, there is no saying how the decision may go; and it is needless to observe how inconvenient it would be to our country if the meridian to which our large mercantile marine have been so long used were changed. Nor would the inconvenience be confined to our country, for many foreign nations as well as our own have adopted the meridian of Greenwich."[90]

The Royal Society had done a volte-face. Its members now considered Britain more interested than any other nation in a prime meridian, a project they thought impractical only five years earlier. William Christie, the new astronomer royal at Greenwich, was very keen on having his observatory chosen as the world's prime meridian. He had attended Rome on behalf of the SAD and was at the Royal Society Council's meeting that drafted the response to the Treasury.[91] However, the Society was backing a prime meridian for navigation at sea, not necessarily for keeping civil time. Free from the baggage of Fleming's twenty-four-hour clock system and zone time, a prime meridian on British soil was an attractive prospect. The sort of conference the Royal Society sought was very different from the one Fleming envisaged. If Fleming was to be a delegate, would he fall in line with the other British delegates?

In February 1884, no one asked that question. The Treasury at last agreed on 13 February to cover expenses for two delegates to the IMC, but left the selection of delegates open for future discussion.[92] It is here that cracks between departments began to widen. The SAD, having organized the British delegation to the Rome Conference, felt that it ought to be in charge of Washington as well. However, the FO was involved because Washington, unlike Rome, was a diplomatic meeting, not just a gathering of scientists. In an attempt to solidify its position, the SAD asked the FO for clarification as to who was in charge.[93] The FO was somewhat relieved to let go of responsibility. In this new arrangement, its only role was to provide the delegates with letters of introduction from Lord Granville, to make their diplomatic status official.[94]

But the FO remained wary of the SAD encroaching on diplomatic affairs. In a private note between FO civil servants, one official wrote that the Greenwich Observatory, the Hydrographic Department of the Royal Navy, or the Royal Society should name the delegates – a regular diplomat would not be much help at the conference, given the subject. But above all, "South Kensington [the SAD] should be shunned (it is no business of theirs, tho' they are trying to push themselves into it)."[95] Unfortunately for the FO officials, they could not beat precedent, and the SAD had the strongest case, having chosen the Rome delegates.[96] The Treasury made the final decision in March 1884: following the practice of the Rome Conference, it put the SAD in charge. The FO would furnish the delegates with credentials and pay their expenses, with funds provided by the Treasury.[97]

In April and May, after the question of expenses was worked out, the SAD set to work selecting delegates. It settled on Captain Sir Frederick Evans, FRS, FRAS, FRGS, the hydrographer of the Royal Navy, and Cambridge astronomer Professor John Couch Adams, FRS, FRSE, FRAS (of Neptune-controversy fame).[98] Alongside these two, the SAD suggested to the CO that perhaps Canada and the Australasian colonies (New Zealand and the colonies – later states – in Australia) might like to send delegates as well (paid for by their governments). The CO ran the idea by the FO, but with concerns. Until this point, no one had thought about a delegate from Australasia. Canada was being considered only because of Fleming's activism. In the practice of fairness, inviting the Australasian colonies was a good idea. But it complicated the diplomatic process. The United States had invited three British delegates, maximum. With Fleming, Evans, and Adams already chosen, there was no room for a fourth. Would the Australasians even be interested? While the CO deliberated, it sent a request to the FO to check if the host might agree to an additional delegate.[99]

The waters were muddied further when, a week later, the SAD announced that the Indian Council had nominated a delegate as well, General Sir Richard Strachey, FRS, FRGS, senior Indian administrator and scientific polymath.[100] The SAD was either unaware of the three-delegate limit or assumed that colonial representatives did not count as British delegates. Strachey had been involved with the Rome Conference, and the SAD was simply following that precedent. Now, however, there were five prospective delegates, with room for only three.

The FO dutifully sent a message to Washington, asking if India, Canada, and the Australasian colonies could be represented. Secretary of State Frelinghuysen wrote back with a reasonable compromise. It would be unfair, he said, if Britain and its possessions had five votes, while each other nation had three. However, involving as many of the world's geographies as possible in the conference would "add to the interest and value of its deliberations and to the weight of its conclusions."[101] Therefore, the United States would allow five delegates from Britain and its possessions, but only three of them would be allowed to vote.

From the U.S. perspective, the offer was fair. But the British had caught themselves in a trap, which only the CO seemed to recognize. Fleming would expect a vote, having been involved in the process since the very beginning. But the Australasian colonies would resent their delegate's having an inferior position to that of Canada's. Seemingly unaware of the nomination of a delegate for India, the CO brainstormed two courses of action: replace one of the British delegates with an Australasian one or simply not tell the Australasians about the conference at all. As of yet, no one had actually approached them; the SAD had merely proposed their involvement. Besides, said one civil servant, Australia did not even have a single railway running east to west yet: would its inhabitants care about the prime meridian?[102] They had not been at Rome, and the colonial secretary, the Earl of Derby, suggested saying nothing.[103] However, some of his staff disagreed, observing that "this office has been more and more in the bad books of the Australian Colonies since the New Guinea question came to the fore, and they might think it a new slight proceeding from Downing Street if they were not consulted."[104] In the end, thinking it unwise to leave Australasia unasked, the CO asked South Australia to communicate with its neighbouring colonies as to whether they would like to send a shared delegate.

The SAD, meanwhile, unmindful of any possible colonial jealousies, decided on its own that the three voting delegates would be Adams, Evans, and Strachey.[105] Its reasoning: those three would receive their instructions directly from the British government, whereas Fleming's instructions from Ottawa might differ.[106]

CO officials were not pleased. They complained to the FO that the SAD's plan was unfair to Fleming, who had been in the mix since the start, and his exclusion might arouse some anger on the part of Canada.[107] In reply, the FO washed its hands of the problem, telling the CO that the decision rested entirely with the SAD.[108]

CO officials were livid. In a flurry of memos, CO civil servants critiqued the SAD's decisions. The SAD, one said, "had no right to appropriate the votes without previous consultation with us, and I think they ought to be remonstrated with, for there is sure to be unpleasantness if Mr. S. Fleming does not get a vote. Could not private pressure be brought to bear on them?"[109] Another CO official went further: "They ought to treat Canada as a Dominion, as she is, and I cannot help thinking that a little pressure brought to bear upon the U.S. Government would make them rescind the decision [to limit the number of votes to three] ... and give one more vote."[110] Yet another suggested writing to the SAD in order to "point out that the unfortunate result of their acting without referring to this Dept. will be to exclude Sandford Fleming who is the originator of the whole Conference, and ask that he may be allowed to have one of the votes."[111]

Before deciding on any course of action, the CO attempted to verify the claim that Fleming instigated the conclave, as it would need firm evidence to justify its complaint about his exclusion. Donnelly (the SAD secretary) seemed to think that the idea had originated in the United States. "It is not certain," one official wrote, "that India would give way to Canada so we must be quite sure of our ground."[112] In response, a full list of Fleming's actions was compiled. Fleming was the first to bring the topic to the attention of the British government back in 1879, although back then the replies from the "English learned bodies were almost entirely unfavourable to the scheme proposed – which shows that the idea had not then commended itself to our scientific men ... The first intimation we had of the Washington Conference was the request from Canada that Mr. Fleming should represent the Dominion there."[113] The note concluded, "But for Mr. S. Fleming the question (as far as I can judge) might never yet have attained official prominence – there might have been no Washington Conference – we might not have been sending any representatives of this country at all."[114]

In Canada, Fleming was unaware of this imbroglio. In June 1884, not having heard anything on the subject officially since he accepted the provisional nomination from the governor general twelve months earlier, he wrote to Tupper and Canadian Undersecretary of State G. Powell, asking for an update. He wanted to ensure his proper accreditation.[115] Powell replied a few weeks later that his office would pass his name on to Her Majesty's Government as Canada's official representative.[116] Powell did not tell the whole story. Tupper informed Fleming that, although he would almost certainly be representing Canada, his voting power was unclear. "The Colonial Office is making a hard fight for you not only on the right of Canada but on your own as the projector of the whole thing."[117]

In the meantime, the CO had heard back from the Australasian colonies. As their delegate would not be able to vote, all of them except Western Australia declined to send a delegate, suggesting that letters from each government would convey their stance on a prime meridian better than a shared, non-voting representative.[118] The CO was relieved, let off the hook for part of the problem. (The CO was taking pains to not cause further insult to the Australasians, but privately denounced their governments as "thin skinned" or "skinless" administrators who detected "slights" where none was intended. It was pleased that Australasia was not interested in the IMC.)[119] At last, only the issue of Canada's vote remained.

Finally responding to the CO's laundry list of complaints, the SAD claimed ignorance of any desire by Canada to participate.[120] It justified excluding the dominion from voting by saying that "the interests of Canada are not necessarily identical with those of this country on the question."[121] As a result, it suggested pleading once more with the United States to recognize a Canadian delegation in its own right, entirely separate from the British.[122] The CO supported the idea as a last resort, and asked the FO to make the request.

It was a long shot. It would mean recognizing a dominion as an equal at a table of nations. This was problematic because most countries would consider the idea a British ploy for an extra vote, assuming Canada would inevitably vote the same way. It would set a "dangerous" diplomatic precedent.

Once again, the United States proposed a compromise. "No colony will be separately recognized," the reply stated bluntly.[123] However, each guest nation could now send five voting delegates rather than three.[124] The compromise was satisfying. As Mr Lowell, a member of Britain's diplomatic mission in Washington, wrote, the new limit allowed "for the representation of the diverse local interests, which might exist within the dominions of any of the powers ... H.M.G. [Her Majesty's Government] can secure representation for all the diversified interests of the British Empire without violating the principle of giving all the powers an equal voice."[125] Britain could now send Fleming, Strachey, and an Australasian delegate with full voting powers, alongside Adams and Evans from Britain itself.

Relieved, the CO quickly ensured Fleming's place. It worried that it might fail to "get the Science and Art Dept. not to pounce upon the additional two votes without consulting this office."[126] A CO official wrote to the SAD immediately to prevent this.[127] Frustrations ran high. Upon receiving from the SAD a redundant note about Canada's voting rights, one CO clerk scribbled across it, "More stupidity!"[128] Another wrote that it would be "difficult to jumble up more mistakes in so short a correspondence."[129] They had cause to regard the SAD with disdain for incompetence, but that is not the entire story. The SAD was populated by members of the scientific community that had long been ignoring and rejecting Fleming's proposals for global time reform. For the CO, the SAD's lack of concern for Fleming's nomination was a sign of incompetence. Seen from another angle, however, the SAD's actions were nothing if not consistent. Fleming's ideas had been found wanting, and therefore it made sense to have Britain represented by its own professional scientists, rather than by this colonial engineer.

The SAD's perspective has thus far been largely missing from this narrative, and the allegations of ineptitude came mainly from the other departments. The SAD's actions in this story remain imperfectly understood because many of its archival records were lost or destroyed. However, it is possible to piece together from private papers some of what was going on there in 1884, enough to partially rehabilitate the department's image. While coordinating the delegates for the IMC, the SAD was simultaneously attempting to prevent an international incident over the metric system.

The "crisis" began in January 1883, when Dr Adolphe Hirsch of Switzerland, secretary of the Metric Bureau, heard about the proposed Washington Conference, and it gave him an idea.[130] He wrote to several prominent British scientists, suggesting that without their nation's participation it would be unlikely that a common prime meridian would ever be agreed upon. However, if Britain participated, and made some concessions, it would be almost certain that it would win out. Accordingly, he asked, could these scientists suggest to their government a barter – accept the metric system in return for France's adherence to a Greenwich meridian. Hirsch laid out a plan, writing that "the experience gained when reforming the weights and measures teaches us unmistakably that a diplomatic conference, as contemplated in Washington, can only succeed provided a basis has been prepared for it by discussion in some scientific assembly, having an official character, so as to settle the scientific questions involved in the matter. In this way the Conference Diplomatique of 1875 was based upon the results arrived at by the Commission International du Metre of 1872."[131]

Hirsch believed the Rome Conference could be a testing ground for Washington, setting out the scientific basis for the diplomatic agreements at the IMC. Such a predetermined agreement was necessary, Hirsch thought, because Europe must go to the IMC united, otherwise the Americans might force the meridian of Washington on the world instead (this was months before the American railways, led by Allen, based their standard time system on Greenwich). With its own major observatory, Washington had a claim on the prime meridian, just as Greenwich or Paris did.

Hirsch's plan started out quite successfully. The Rome Conference did suggest some sort of Franco-British entente over the metric system and the Greenwich meridian, albeit watered down significantly by the British participants. Christie, Strachey, and the Rome delegates appointed by the SAD wanted Britain to join the Metre Convention and pay dues to it, but not necessarily begin using the metre. They recognized that the Convention ensured accurate conversions between imperial and metric units, so Britain ought to pay its share for that service.

Even this lesser bargain, however, was fraught. In Britain, staunch supporters of the imperial system like Smyth regarded such a concession as a slippery slope and worked to undermine any such agreement. H.J. Chaney, the British warden of the standards, firmly opposed the idea and discouraged the Treasury from offering any such payment. But international pressure was rising. Britain was not pulling its weight; it benefitted from the Convention's conversion tools but did not pay dues.

The SAD was caught in the middle of this crisis when the issue of nominating delegates for Washington came up in early 1884. The SAD had to ensure that any delegates chosen would have moderate views about the Metre Convention. Fleming was a wild card where it was concerned. It wanted it own people in the delegation.

This was no trifling disagreement. The Treasury, backed by the indignant Chaney, resisted payment to the Convention, and offered only a token amount. Wilhelm Foerster, a German member of the Convention, explained to Christie that such a gesture was an insult: "You regard such payments as 'a step gained' and I have no doubt that you are right, and that this offer will be only a transition to a full adhesion of your government to the Metric Convention. But this transitory step ... was necessarily in our eyes of a nearly offensive character."[132] According to

David Gill, the Scottish-born chief astronomer of Cape Colony in what is now South Africa, "The proposed terms have been received with a very strong feeling of indignation."[133] Other countries paid far more for the same benefits, and some of these were already inclined to dislike Britain. It was being asked to pay the same amount as Turkey, and far less than the United States. Neither of those countries used the metric system, yet they paid their share. Britain's refusal to pay, wrote Christie, "seems likely to raise a storm of indignation which will put us in a position of complete scientific isolation."[134]

Christie set about lobbying the Treasury to make the payments. He began by asking a Liberal politician, Peter MacLiver, to raise the matter in Parliament, hoping it might "quicken the movement of the Treasury."[135] With MacLiver's help, a deputation was arranged to visit the chancellor of the exchequer. The meeting seemed to go badly. Chaney spoke against the idea, and Christie feared that they had failed to make themselves "intelligible."[136] But the response was actually somewhat positive. The Treasury explained that, before the deputation, it had been unwilling to pay anything that moved Britain towards the metric system, as it was "repugnant to public feeling."[137] It had, however, been considering the subject with due diligence, understanding that the IMC decision on a Greenwich prime meridian might "be influenced by the action which may be taken by the United Kingdom" on the metric question.[138] But the two subjects had seemed so unrelated to the Treasury that it deemed the cost not worth the possible gain. When the deputation pointed out that joining the Metre Convention of 1875 did not necessitate using the metric system, and that the costs were lower than the Treasury had believed, the Treasury was more willing to consider the request.

In the meantime (the spring of 1884), the SAD was in the midst of its messy negotiations with the CO and the FO over the Washington delegates. Donnelly managed to secure assurances that the SAD would be picking delegates, and it chose people who would support the Metre Convention in return for the Greenwich meridian. As Strachey was its ally in this, Christie suggested that Donnelly ask the Indian Office to get him appointed as an Indian delegate.[139] Adams and Evans were also Christie's idea.[140]

The SAD's outlook on the Washington Conference was based on solving an entirely separate scientific controversy, namely adherence to the metric system. Fleming's timekeeping schemes were immaterial, so the SAD ignored him. Even had it known about his desire to attend Washington, his metric ambivalence disqualified him in the SAD's eyes (i.e., his having ties with both sides, via metric-supporting Barnard on one hand and the IPAWM on the other). He was therefore nearly passed over, but for the CO's intervention. Thus the result of this controversy was that most of the British delegates were picked for their stance on the metric system, not on the meridian. Ironically, the tenuous link between the two concepts was deteriorating rapidly and would evaporate entirely by the time the conference began. The metric system would be discussed only perfunctorily at the IMC. On both sides of the English Channel, the idea of a quid pro quo, trading the Greenwich meridian for the metric system, had lost favour.

French politicians decided that they would not support a Greenwich meridian no matter what concessions Britain made. As Hirsch put it, even politicians who privately supported a common prime meridian would not dare be seen accepting Greenwich over Paris. Public opinion was fervently anti-British in 1884, due in large part to Britain's recent military interventions in Egypt.[141] On the British side, the trade was never going to work either. Although the Treasury was now considering paying its share to the Metre Convention, it had taken too long, and it was clear that the matter was to be considered separately from the meridian. David Gill told Christie that the two matters had to be kept separate going forward. Even if France did not accept Greenwich now, Gill said, eventually international pressure would probably compel it to do so.[142] Christie wrote apologetically to Hirsch: "Official matters take a long time in this country when several departments of the government are concerned and the question has to be argued out between them by correspondence ... I think that this question of adhesion to the Metric Convention will be considered entirely on its own merits quite apart from the adoption of a prime meridian. The feeling, as far as I can judge, is decidedly against making any bargain with the French or anybody else to adopt the metrical system if they will adopt the meridian of Greenwich."[143]

It was July 1884 before the Treasury made a final decision – to join the Metre Convention, but without paying arrears.[144] Christie lamented the long, messy process: "It is unfortunate that our government was so slow to be convinced of the desirability of this step, and that thus the good effect which it might have had on the settlement of the Prime Meridian question has been lost."[145] By the beginning of the Washington Conference in October, the two questions could not be further apart. The SAD, which had tried so hard to appoint metric advocates, had to tell the delegates not to raise the topic at Washington. Its official letters of instruction to them reiterated that point.[146]

During the key phases of delegate selection, the metric controversy dominated the SAD's attention. Its resulting inattentiveness to the other

departments nearly barred Fleming from attendance. Ultimately, the
CO's campaign to secure him a spot paid off, and Fleming was seated
alongside Adams, Evans, and Strachey. But it was a close thing.

The actions of the SAD, the Royal Society (represented by Christie),
and their chosen delegates paint a stark image of an insular scientific
community fixated on its own goals and priorities, all quite different
from Fleming's. Even when they finally abandoned the plan to adopt
the metric system in exchange for the Greenwich meridian, their interest in timekeeping was as a tool for navigation and astronomy, not as
a revolution in public timekeeping practices. They were not interested
in Fleming's notion of universal, collective timekeeping. Fleming and
his fellow British delegates may have been sitting at the same table in
Washington, under the same flag, but they had very different ideas about
what was to be accomplished there.

LOOSE ENDS AND LAST-MINUTE ARRANGEMENTS

Fleming received his official appointment to the IMC on 25 September
1884, just six days before it was to begin.[147] As for Australasia, the question of its attendance was reopened in September. Although these colonies
had already declined to send delegates, and had offered to send letters
instead, the situation had since changed. Previously, they believed they
would have no vote. Now, they could have one, so attendance seemed
much more palatable. With the conference less than a month away,
could a delegate arrive in Washington in time from the South Pacific?
Or were there any sufficiently qualified Australians or New Zealanders
already nearby to represent them? James Cockle, a mathematician and
the former chief justice of Queensland, was chosen; he was in Britain,
and therefore had a more manageable journey to Washington.[148] All the
Australasian colonies except for Western Australia agreed to pay their
portion of Cockle's expenses.[149] However, by the time he was informed,
it was too late, so he declined the nomination.[150] The United States did
not find out there would be no Australasian delegate until 2 October, by
which time the conference was already in session.[151]

Fleming's official title at the IMC was "British Delegate for Canada."
Under the agreed-upon provisions, he could vote alongside Adams,
Evans, and Strachey as a representative of the British Empire. But one
final hiccough undid all of that work. Despite the months of incessant
correspondence and quibbles about voting, the conference itself decided
on the spot in October that votes would be cast by nations, not by

individuals. And since Fleming was representing Britain, not Canada independently, his voice was significantly diluted.

His American friends fared almost as poorly. At first, all seemed well. Fleming's ally Barnard, who was president of the AMS, was chosen to lead the American delegation. Commander William Sampson, a staff member at the U.S. Naval Observatory in Washington, DC, was second. But American scientific periodicals lambasted these choices, Barnard for being too old (he was seventy-five) and deaf (he had been deaf since a young adult due to a congenital condition and had been a teacher of the deaf); Sampson for lack of scientific credentials.[152] While Fleming had been nearly excluded for his colonial status and his engineering profession, inexperienced Sampson lacked prestige. Barnard, meanwhile, was well known and respected, but critics felt his old age and disability had no place in the argumentative, masculine worlds of science and diplomacy. He had other enemies too, though less reputable. Fleming's unfortunate anti-metric acquaintance Charles Latimer was particularly incensed at Barnard's selection. In July 1884 the IPAWM wrote a lengthy diatribe to the president of the United States, Chester A. Arthur, exposing the supposed pro-metric conspiracy hidden beneath the conference's cover. It claimed that twenty of the thirty nations attending were metric supporters, and pointed out that even the leading American delegate, Dr Barnard, was favourable to it. The letter went on to suggest a member of the IPAWM be allowed to attend, to counterbalance the misguided metric reformers.[153]

The IPAWM's efforts were to no avail, so Latimer wrote to Fleming in August: "I hope the Canadian Government will appoint you and I beg that when you get to Washington that you will guard the question of weights and measures and let it not come up at all. Mr. Barnard's whole object is to bring up that subject as you will find when you come to the convention."[154] Despite Latimer's fears, Barnard was interested in time reform quite independently of his metric sympathies. In late September Latimer wrote to Fleming again, suggesting that the conference "is all worked up for the Metric System by Dr. Barnard ... I think it is for the purpose of inveighling [sic] the United States and Great Britain to adopt the French Metric System and you will find that everything they do and say will tend in that direction. Watch it carefully I beg of you."[155] Latimer's fears had some grounds, as we have seen. The idea of a Greenwich-metric trade had indeed been discussed at Rome, but Latimer overestimated the danger. By the time of the Washington Conference, the idea had been soundly rejected.

ASCE engineers wanted to be represented at Washington too. John Bogart, who had worked with Fleming in the ASCE, was upset that U.S. "engineering and transportation interests" had no spokespeople.[156] His preferred candidate, a railway engineer named Mr Whittemore, was not picked, so he was relieved when Fleming (though not an American) had secured his place, and even more pleased when William Allen, the architect of standard railway time in North America, was picked as a third U.S. delegate.[157]

The final two American delegates were Cleveland Abbe (a rare victory for Fleming) and lawyer and amateur astronomer Lewis Rutherfurd. Going into the conference, it seemed a strong position for the time reformers, with Fleming, Abbe, and Barnard all attending. But just days before the event began, Barnard, who was supposed to be in charge of the U.S. delegation, dropped out under pressure over his disability. He wrote to Fleming that on account of poor hearing he decided to give up his seat to Simon Newcomb or Eugene Hilgard, both eminent scientists.[158] Although he gave up his seat by choice, the outcry in scientific journals about his hearing impairment influenced him. What might have been a capstone to a long career instead became a minor scandal.

One of the French delegates, Jules Janssen, visited Barnard in September just before the conference. Janssen wrote with regret that Barnard would not attend, calling him "a very deaf, wise old man (we could not talk with him without an ear trumpet), but full of the serenity that a long, interesting career gives."[159] The Frenchman regretted that this *éminence* had stepped down in favour of "an ordinary naval officer," and seemed to imply that Barnard would have been more attentive to French arguments (a fair assumption, given his support for the metric system).[160] His replacement, Admiral C.R.P. Rodgers, retired superintendent of the Naval Academy, was an unknown. Hilgard and Newcomb, whom Barnard suggested as replacements, were not invited. When Rutherfurd had been appointed, he wrote to Newcomb, "I do not know why I was named … [and am at a loss] to find a good reason why he [the president] has not appointed yourself and Hilgard both of which nominations would have been most manifestly proper."[161] Nonetheless, the final U.S. spot was given to Rodgers. Previously ambivalent about time reform, Rodgers acted as a somewhat neutral party, replacing Barnard as head of the American delegation. When the conference began, he was elected its president.

With the pieces finally in place, the IMC in Washington could begin. But as we have seen, the months and years prior can tell us a lot about the process of scientific and diplomatic decision-making. Competing interests, sometimes only tangentially associated with timekeeping, shaped the discussion. Significant opposition greeted Fleming's attempts to bring universal public timekeeping before the world's scientific community, and the concept progressed only by means of relentless lobbying and through some chance developments. Fleming was an outsider. His position as an engineer rather than a scientist, alongside his colonial status, severely limited his scope of action. His early papers and presentations were ignored and derided, and he struggled to find legitimate allies, therefore seeking allies also on the fringes of the scientific community, like the IPAWM. At the same time, rivalries among government departments, and international convention, nearly excluded him politically. Through incessant, tireless efforts, he ensured that his ideas were made known, but they were rarely listened to, except by his peers in the railway profession, like William Allen. In the world of science, he was a nobody.

This conclusion requires some caveats. It would be incorrect, for example, to regard the scientific establishment as a nasty hegemon, a foil to Fleming's underdog heroism. The notion that European scientists were a conservative elite, holding back the tide of progress, with Fleming as an embattled, enlightened hero, is a false one. Such a judgment would, at best, be a new exercise in myth-making.

A more accurate picture would recognize that there existed multiple, variable, and competitive opinions on timekeeping in a period when the question was very much up in the air. Nothing about Fleming's plan was inevitable or natural. His time-zone idea, eventually adopted, could have been replaced by a very different system, and in fact for years his position was the minority one. The establishment position held by men like Christie and Airy was that universal time was a tool for scientific research, not a public good. Most of the delegates to the Washington Conference attended not to discuss public timekeeping, but rather to establish a meridian for navigation and astronomy: in other words, universal time was to be a specialized tool for specific professions, not, as Fleming wished, a revolution in civil time and in public behaviour. Opinions on time reform were deeply held. These differences in opinion, and the players' similarly differing goals, shaped the debate. Even before the conference began, these motivations simmered beneath the surface, moulding perspectives and deciding, ultimately, which ideas, and which people, had a legitimate seat at the table, both literally and figuratively. Debate and discussion shaped what was possible. Time reform was a matter of negotiation, not of invention.

2

Amateurs, Professionals, and Eccentrics

Between 1876 and 1884, while railway engineers such as Sandford Fleming and William Allen were developing their idea for a new universal system of timekeeping, a second, very different vision of universal time was emerging within the astronomical community. Unlike the engineers, astronomers had no desire to force the public to change its timekeeping behaviour. Instead, they wanted to create a system of universal time as a specialized tool for astronomical observations – so two astronomers observing the sky from different places on earth could use the same time system. The public need not use or even know about this time system, as far as they were concerned.

At the International Meridian Conference (IMC) in Washington, DC, in the autumn of 1884, these two very different conceptions of universal time would clash. This disagreement – was time a specialized tool or a public good? – would shape the outcome of the gathering. Indeed, it would shape the very future of the global time system. But in history, as in life, the full story is always messier than any simple two-sided conflict would suggest. These two visions of time reform may have been the dominant ones, but they were constantly being challenged by other ideas. So while chapter 1 examined the origins of railway standard time, and this one begins by telling the story of the other side, of astronomical universal time, it also discusses how outside forces challenged and refined those two dominant conceptions of global timekeeping. Time reform was influenced not just by science and technology, but also by religion and society and class and culture. Participants in the time debates approached the subject from a multitude of backgrounds, beliefs, and skillsets. These disparate perspectives lent wide-ranging breadth and depth to the time debates. Time reform intersected with, and was shaped by, dozens of other movements, discoveries, and controversies under scrutiny during the late nineteenth century.

These diverse approaches to time reform filtered through, in varying degree, to the professional astronomers, engineers, and naval officials who made the decisions at the IMC. We have already seen how the metric debate shaped the approach of Britain's Science and Art Department (SAD) to the conference. Similarly, we've seen how Fleming's North American campaign for a new universal system of time zones united railway authorities, yet failed to convince astronomers. But contemporary debates about religion, international politics, and even archaeology also affected the buildup to Washington. Timekeeping practices were social and cultural products, as well as political and scientific ones, with multiple and varied influences.

Similarly influential was the ever-changing relationship between amateurs and professionals. These relationships determined whose voices were heard and whose were not, and thus which of the various intersecting concerns most influenced time reform. Professionals with the weight of academic authority behind them commanded more sway than those without it. Late-nineteenth-century professional science, especially in astronomy, was insular, with little space for dialogue with non-specialists and amateurs. This led to a particularly narrow view of time standardization, which proved incompatible with Fleming's sweeping proposals for civic time reform.

Of course, the notion of insiders and outsiders is overly simplistic. In reality, the boundaries of scientific networks were shifting and flexible. People moved in and out of the astronomical community, and circumstance and personal relationships radically altered people's status within it. Its boundaries were permeable, but they were very real, and the battle lines over time reform followed professional more than national boundaries.

To explore these issues, this chapter has four parts. The first describes the predominant vision of universal time among astronomers, and how it came about, mostly because of the greatest astronomical undertaking of the century – the observation of the transit of Venus across the sun. The other three parts of the chapter provide conflicting or contextualizing perspectives on time reform from the periphery of the astronomical community. Each traces the career of an individual who personifies a set of beliefs or debates that intersected with time reform. They ranged from a respectable scholar, through a victim of professional discrimination, to a purveyor of intellectual snake oil. Each case reflects the cultural and intellectual milieu of Victorian professional science, in which the time-reform debates were embedded and that forged the proposals and decisions made about time at the IMC (and beyond).

The first case discusses Victorian faith as an informant of the time debates, as seen through the works of Irish-British astronomer Annie Russell, and her husband, with whom she collaborated. The second case follows English navigator William Parker Snow's hapless attempt to inject the language of nautical safety into the time-reform debate. The final case uses the career of Scotland's astronomer royal, Charles Piazzi Smyth, to examine the intersection of time reform with atypical religious belief, archaeology, nationalism, and the Victorian fascination with Egypt. Built into all these parts is the constant underlying tension between professionals and amateurs. Together, these three case studies paint a fuller picture of their world, in which timekeeping was messy and incomplete and open for reinvention, by professional scientists and amateurs alike.

NINETEENTH-CENTURY ASTRONOMY

Professional astronomy in the late nineteenth century boasted a global network of scientists, yet it also remained a tiny, tight-knit, insular body with a fixed conception of time measurement. Of all the scientific disciplines, astronomers had the most obvious connection to time reform, because their observations both depended on accurate timekeeping and conversely produced it for others who might make use of it. Geographers were similarly interested in time, as measuring it could help determine longitude and facilitate land surveys. Both mapping and navigation relied on it (and thus so did imperial conquest, regulation, and taxation: hence this was not merely an academic pursuit). In effect, scientific disciplines that involved observing the natural world on a grand scale required some way to standardize time. In contrast, disciplines that examined the microscopic, such as chemistry or botany, though perhaps having to keep time accurately, did not need their clocks standardized worldwide. Astronomers and geographers (and their nautical counterparts) were therefore the main scientific players in the debates over time reform.

In the 1870s and 1880s, one rare phenomenon dominated the attention of the astronomical world and would come to shape its understanding of time and its measurement. The transit of Venus across the sun occurred only twice in the nineteenth century: in 1874 and 1882. Such an event would not occur again until 2004, so these rarities spurred some of the era's largest astronomical undertakings: a global effort to observe the transits and, from the data gathered, determine the size of the solar system. By observing the transit, astronomers hoped to be able to calculate the astronomical unit (AU), or the distance between the sun and the earth. Observation points were set up in dozens of locations, with experts travelling across the globe to witness the event. To determine the AU required observations at different points around the world for comparison by triangulation. Accurate, standardized timekeeping was essential to properly compare the observations at each location. More than this, the exact longitude of each observation point relative to Greenwich (or another national observatory) had to be determined, a delicate and arduous task requiring expensive, accurate chronometers, sophisticated astronomical equipment, and months of observations at each site.[1]

So intrinsic was time standardization to the transit-of-Venus observations that it was these expeditions that most astronomers had in mind when they considered time reform in the 1880s. In their minds, standardization was tied inextricably to the precision and accuracy of transit calculations, and such perfect standardization was a much more challenging task than coordinating the less-accurate clock time required for civilians. This is why, in large part, Fleming's time reform for the general public seemed so far-fetched to astronomers. Standardized time, for them, was a task suited to scientific endeavours, a specialized tool irrelevant to the average man or woman.

The transits of Venus were observed globally. While Europeans led many of these expeditions, astronomers from every continent participated. Latin American astronomers, like Luis Cruls in Brazil and Angel Anguiano in Mexico, shared their observations with European and Japanese counterparts, and vice versa. Unfortunately, in both 1874 and 1882, inclement weather ruined the observations at several stations around the world, interfering with the results. But the effort and coordination required reveal the broad geographical scope of participants.

Observing the transit of Venus was both a cooperative international effort and a competitive one. Nationality divided the community of astronomers, but engendered rivalry more than seclusion. By necessity these events required cooperation, and data and techniques were shared across national borders.[2] Language differences, though impeding cooperation, did not tend to divide astronomers into "silos" of English, French, German, and so on. Rather, language limited the class of people able to participate in the international dialogue to those with a second or even third language (usually French). In Britain, for example, most children were not taught French in the Victorian period. The Elementary Education Act of 1870 required the teaching of reading, writing, and arithmetic (and needlework for girls) in elementary school, but not foreign languages.[3] Only middle- and upper-class children were likely to attend post-elementary school, and they too were subdivided. The Taunton Report of 1868 separated secondary schools into three tiers: a first-grade liberal education for those going on to university; a second-grade education, including two modern languages and Latin, for those going into the civil service and army; and third-grade schools that taught some basic French and Latin for likely tradespeople.[4] Curriculum was thus divided by class, and by gender, limiting the opportunities available to working-class men and women.[5] In this way, language was a barrier not so much between nations as between classes. Scientists tended to be drawn from the middle and upper classes, creating a transnational network of privileged intellectuals. Societies such as the Royal Society and the British Association for the Advancement of Science (with its local chapters) were connected by large international gatherings, which acted as the lifeblood of these networks. Mobility, and the resources to attend these gatherings, were almost a necessity for active membership.[6]

Within the scientific network, some institutions, societies, and individuals were more influential than others. Networks, as historian Frederick Cooper so vividly describes, are "filled with lumps, places where power coalesces surrounded by those where it does not, places where social relations become dense amid others that are diffuse."[7] This was very clearly true of the world's astronomical community of the 1870s and '80s, where personal connections and systematic forms of marginalization by race, class, and gender shaped so much.[8] The astronomical community spanned the globe, but their cohort was limited in other ways.

In this global network of unevenly distributed nodes of power, connected by a dialogue of ideas, discoveries, and challenges, and channeled through institutions, which ideas and people gained acceptance? The answer is intersectional. Some methods of exclusion seem obvious – lack of education and multilingualism, and therefore class, blocked many. Similarly for race: in Britain, for example, conceptions of race were hierarchical and exclusionary, based in theories of Social Darwinism, which lent them legitimacy. Phrenology and similarly problematic anthropological studies shaped public understandings of the colonial "other." Colonial subjects visiting Britain were subjected to various challenges, but their experiences were varied, shaped by their own circumstances. Visitors from India, for example, ranged from destitute workers to honoured guests of royalty, depending on their station.[9] Some made careers in medicine, law, and science, either educated in Britain or at home. Britain in the nineteenth century could even be a "site of pleasure and advancement" for Indian aristocrats.[10] But these were the minority. Most colonial visitors to the imperial centre found themselves outcasts.

Racial difference was similarly divisive beyond the empire. Britons saw themselves as locked in a fierce competition with other races, fueling Anglo-German, Anglo-Russian, and Anglo-French antagonism, and heightening tensions with "internal others," such as Irish nationalists.[11] But even within this framework of racial competitiveness, boundaries were occasionally crossed. The Japanese delegate at the IMC in 1884, Kikuchi Dairoku, for example, though neither British nor a colonial subject, was educated in Britain. He was the first Japanese graduate of Cambridge University and attended lectures with renowned physicist Lord Kelvin in the United States. He later became president of Tokyo Imperial University. Very clearly belonging to the international scientific community, Kikuchi demonstrates that racial bias could be overcome in the right circumstances. Just as "the gendered body is performative," as Judith Butler argues, so is the racialized body.[12] Non-European scientists like Kikuchi adopted western modes of dress, for example, in order to more easily find acceptance in Europe. Many Japanese elites similarly donned western dress *en masse* as part of a policy of "modernization" adopted by Japan after the Meiji restoration in 1868.[13] Imitation was often the price for inclusion in the international community.

Gender could equally hamper participation. Given the nineteenth-century European middle class's tendency to separate men and women into public and private spheres of influence, respectively, women had difficulty maintaining a public position in any scientific network.[14] But, as always, these rules were incomplete and could be undermined. To begin with, the notion of separate public and private spheres was an ideal, rather than a reality.[15] Women often participated in the public sphere, by choice or necessity, and many found roles in the sciences as well. Women who expressed their scientific interests in terms of "natural theology and thereby moral education were acceptable in scientific narrative."[16] Female botanists, for example, studied God's creations, "performing" their expected roles as religious and moral wellsprings through the study of the natural world, and in doing so found a way to gain acceptance as practitioners of science.[17] Women made inroads in other fields as well, and after mid-century higher education for upper- and middle-class women slowly became more accessible, although it was "heavily contested."[18]

To be an astronomer in the nineteenth century, then, was to conform to a narrow set of identities. Race, class, and gender acted to exclude outsiders, although performative behaviour allowed certain individuals to overcome these barriers. "Mainstream" astronomy therefore, despite its global reach, was a small and homogeneous community. But its periphery boasted more diversity, and through those boundaries filtered a stream of ideas and beliefs that would alter the community's understanding of timekeeping. If the careful precision required for watching the transit of Venus represented the dominant astronomical take on the uses of time, these peripheral ideas were complicating and enriching that narrow view.

ANNIE RUSSELL

The intersection between timekeeping and Victorian culture cannot be explored without some mention of religion. Religion was central to Victorian life, and its relationship to timekeeping is examined here through the eyes of astronomer Annie Russell. Russell worked at the Royal Observatory, Greenwich, in the 1890s, where she was helping to observe, calculate, and distribute Greenwich time to all of Britain. Her experiences at the observatory, and her publications concerning historical astronomy and the scientific interpretation of the Bible, reflect a practical and grounded approach to timekeeping, tinged with theological implications. Although most of her work was written after the 1884 conference, Russell's perspective is representative of the interactions between astronomy, time reform, and Christianity in late-nineteenth-century Britain.

Religion influenced nearly everything in the Victorian world. At the IMC, for example, proposals for Rome or Jerusalem to mark the prime meridian were voiced with as much rigour as Greenwich, based on religious grounds rather than scientific principles. Russell's career exemplifies this close relationship between science and religion. Her expertise in astronomy and in religious texts showed her that all of humankind's timekeeping methods were ultimately arbitrary. Timekeeping practices had varied widely the world over, from biblical times to the present, so there was no singular, "correct" way to do it. The exception was timekeeping methods founded on natural astronomical rhythms. The smallest possible division of time that could be ascertained, without mechanical timepieces, was the day, determined by the turn of the earth.[19] Smaller divisions like hours and minutes, according to Russell, were superficial additions to the natural truth of God's heavenly creations.

Russell, more than time-reform activists such as Fleming, was able to view the time-reform movement in the perspective of the *longue durée,* an inconsequential blip in the pattern. Timekeeping methods were ultimately predetermined by the stars (the work of God), and were unchangeable at their core, whatever cosmetic finishes reformers might want to put on them. Russell viewed the heavens through the dual lenses of both her telescope and her Bible, a mix of science and religion that was not uncommon in Victorian Britain. Both would shape the era's time-reform debates, at the IMC and beyond.

The relationship between religion and science was a contested subject. For many people, the two were easily compatible. Science, as the successor to the natural philosophy of the eighteenth century, was undertaken to better understand creation, and could therefore be seen as an act of devotion in itself. But there were those who came to see religious teaching as antithetical to science, such as John William Draper, whose *History of the Conflict between Religion and Science* (1874) popularized that conflict.[20] Most scientists fell between the two extremes, suggesting that religious belief was compatible with science, but the worst forms of dogmatic sectarianism were not.[21]

A subsidiary issue in Europe, closely related to science versus religion, was whether biblical texts were literal or metaphorical. If an astronomical observation or scientific discovery contradicted a passage from the Bible, which took precedence? The most famous such debate revolved around Darwin's *On the Origin of Species* (1859), but it was representative of a broader controversy, some of which went to extremes. Literal interpretations of biblical texts led one lecturer named Samuel Rowbotham to conclude that the earth was stationary, flat, and only a few thousand years old. Writing under the pseudonym "Parallax," in 1881 he published *Zetetic Astronomy: The Earth Not a Globe*, which argued for biblical literalism and denounced modern astronomy for misreading the nature of the heavens. Rowbotham and his followers engaged in several high-profile public experiments in an attempt to prove their theories. When these experiments failed, they spun them so as to discredit the experiments' legitimacy and so kept a sizable number of devotees.[22] In one incident in 1870, a Zetetic follower named John Hampden bet £500 to anyone who could prove the globe was round. Alfred Wallace, a naturalist who had independently conceived of natural selection alongside Darwin, took Hampden's bet and won it.

These debates had obvious ramifications for time reform. The calendar was determined by the movement of heavenly bodies, so the nature of that movement was of real concern to time reformers. Fleming's time-zone proposal assumed a spherical earth, for example. The possibility of a flat earth would have undermined his entire project.

The career of Annie Russell (1868–1947) (Figure 2.1) was equally entangled in these questions concerning both scientific and religious

Figure 2.1 | Annie Russell Maunder observing a solar eclipse.

timekeeping. She worked at the literal and symbolic centre of global
timekeeping (Greenwich) and was fascinated by the relationship between
biblical texts and modern astronomical observations. The daughter of
an Irish Presbyterian minister, she won a three-year scholarship in 1886
to Girton College in Cambridge – one of the earliest colleges in Britain
open to women. Founded in 1869, the college offered courses in mathematics, science, and classics but did not confer degrees. Nonetheless,
on completion of her coursework, Russell was able to find a position at
Greenwich Observatory, which she took up in 1891.

   She was one of several women employed there as computers, the others being Alice Everett, a fellow Girton graduate who went on to be
a physicist and engineer, and three women from Newnham College,
Isabella Clemes, Harriet Furniss, and Edith Rix.[23] Women were similarly employed at the Harvard Observatory in the United States at
about the same time, but such appointments for women were rare.[24]
The Greenwich positions, though highly technical, were not particularly prestigious. As computers, these women were doing the grunt

work of astronomy, mostly mathematical calculations and some limited observations with the observatory's instruments.[25] Most of the computing was normally carried out by teenaged boys in temporary positions, who used it as a stepping stone into the civil service.[26] They had only to sit an entrance examination to earn the position, while women had to have studied at a university ladies college.[27] Nonetheless, these women proved more capable than most of the younger boys, and were trusted with more complicated tasks and responsibilities.[28] For example, Everett, Rix, and Russell all took observations with the transit instrument, which established Greenwich time for all of Britain. The time was distributed from the observatory twice daily, immediately after they had taken their observations, when the time signal was most accurate. It was sent by wire to the Post Office headquarters (on St Martin's Le Grand, in the City of London), whence it was disseminated to the rest of the country. These women were providing all standardized time for naval, astronomical, and civil purposes. While working the transit instrument, they were quite literally producing accurate scientific time for Britain and the world.

But Russell's role as timekeeper for the world lasted only four years. There were restrictions in place that kept women out of the observatory in the long run. To begin with, the salary was just £4 a month,[29] or roughly half what Russell had been earning as a schoolmistress.[30] For Russell, the experience and unique opportunity outweighed the low salary. But for other women with fewer means, the low pay was an impassible barrier. Several women, including astronomer Agnes Clerke, turned down the job because of the salary.[31]

Even during her tenure at the observatory, Russell was not entirely welcomed by the scientific profession. Although she was nominated to be a fellow of the Royal Astronomical Society in the early 1890s, her gender excluded her. Two decades later, at the start of the First World War, she was nominated again, but this time she herself declined in protest. She wrote that the society had blackballed her "solely because I was a woman. But – I remain a woman still; a fact I never regretted until the call came for *men* to join Kitchener's Army."[32] Russell's career is an excellent example of the alternative channels of scientific pursuit that were possible in late Victorian Britain. Faced with exclusion, she created her own space for astronomical endeavours.

Nonetheless, Russell faced challenges her male counterparts did not. Only single women could be employed at Greenwich, and upon marriage a woman had to leave.[33] Russell was therefore forced to depart on marrying another Greenwich employee, Walter Maunder, in 1895.[34]

But that was not the end of her career. Russell carried on as a partner in her husband's observations, collaborating with him on several books. She was not alone. Women were often a silent partner in their husband's scholarly work, acting as editor or author, though sometimes published under their partner's names. Others acted as intellectual confidantes.[35] Henriette Janssen, for example, the wife of French astronomer Jules Janssen, was in constant communication with her husband during his astronomical voyages and at the Washington Conference in 1884.[36] The exclusion of women from science therefore was incomplete, and often circumvented by upper- and middle-class women keen to participate.

The intersection of religion, astronomy, and timekeeping is clear in Russell's publications. As we saw above, women were more accepted in scientific professions when their work was tied to morality and religion. They were also more likely to be relegated to amateur status, kept out of official institutions and denied professional membership. Barred from professional societies, Russell embraced both amateurism and religious study.[37]

If amateurs and the public were Russell's audience, her passion was biblical astronomy. Much of her and her husband's work (often published under his name, but their work was shared) concerned either public astronomy, such as her *The Heavens and Their Story* (1908), or the study of ancient texts with reference to modern understandings of the night sky, such as in the co-written *The Astronomy of the Bible*.[38]

Both these books explore human timekeeping, but what they left unsaid speaks volumes. There is little discussion of the prime meridian or standard time or time zones, no analysis of IMC resolutions or the universal day. All of that, for Russell, was cosmetic, the mere trappings of modern timekeeping. For the two angles of her work (amateurism and religion), they were irrelevant. Timekeeping in the Bible was derived from astronomical bodies, not clocks and watches, and the amateur astronomer needed a clock only to better measure the degrees of a circle (Russell illustrates 47 [*sic* 48] degrees as equal to the space a clock hand moves in eight minutes on a round twelve-hour clock face, for example).[39]

Russell prepared *The Heavens and Their Story* as a handy guide to new astronomers. She described how amateurs lacking a telescope could employ Hilly Fields (a park in Lewisham with heights up to 175 feet above sea level; three miles southwest of Greenwich Observatory) as an open-air observatory, using nearby church spires as meridians and points of reference.[40] But Russell was pointing to the heavens in an attempt to teach about creation beyond earth. She spent very little time describing

practical astronomy, e.g., taking transit measurements to determine longitude or time, and more on teaching her readers how to determine the
length of a Martian day.[41] Her goal was to convey "the vastness and
mystery of that great starry system of which our sun and his family
occupy a small and insignificant corner."[42]

In *The Astronomy of the Bible*, Russell and Maunder took a similar
approach, beginning the chapter on timekeeping with the length of the
day on Venus, rather than on earth.[43] But they quickly turned their attention homeward to discuss ancient timekeeping systems on earth, comparing biblical passages to modern observations of the stars. They wrote,
"The making of the calendar is in all nations an astronomical problem:
it is the movements of the various heavenly bodies that give to us our
most natural divisions of time ... but as there are many heavenly bodies
and several natural divisions of time, the calendars in use by different
peoples differ considerably."[44]

Though discussing ancient history and myth, Russell and Maunder
were also commenting on current timekeeping. They thought that
squabbling over whose time was right was unproductive. Any divisions
of the day beyond what could be ascertained by the stars were arbitrary. For example, Russell and Maunder explained that while ancient
Judaea's ecclesiastical day began at sunset (as in modern Jewish obser-
vance), its civil day commenced at dawn. Likewise, "we have a similar
divergence of usage in the case of our civil and astronomical days; the
first beginning at midnight, and the second at the following noon, since
the daylight is the time for work in ordinary business life, but the night
for the astronomers."[45] Differences like these were merely products of
convenience, not worth fighting about. Of course, as we see next chapter, Fleming and other reformers wished to unify time measurement
and wanted astronomers to replace their astronomical day with the
civil day. But to Russell and Maunder, such conflict was unnecessary,
even short-sighted.

Religious justifications for methods of time measurement were common in the timekeeping debates. Some commentators condemned certain
methods of timekeeping by associating them with rival denominations or
sinful practices. But religion led Russell to a far more flexible approach
to the issue. After years of studying both scripture and the stars, she
came to believe that there were only a few natural or God-given measures (the solar day, the lunar month, and the year), and anything else
was either superficial adornment or constructed convenience, so there
was little point in arguing about timekeeping beyond finding what was

most convenient. This perspective was radically tolerant, and made the vitriolic timekeeping debates seem somewhat petty.

Russell was not a biblical literalist: she would have abhorred the opinions of Rowbotham and Hampden. The final chapter of *The Astronomy of the Bible* attempted to explain the star of Bethlehem, but concluded that scripture did not contain enough evidence: "The narrative [in the gospels] appears to me astronomically too incomplete for any astronomical conclusions to be drawn from it. The reticence of the narrative on all points, except those directly relating to our Lord Himself, is an illustration of the truth that the Scriptures were not written to instruct us in astronomy, or in any of the physical sciences, but that we might have eternal life."[46] It was a direct attack on literalists like Rowbotham and Hampden. Russell was open to any convenient form of timekeeping, since God had left no natural clues for its measure beyond the solar day, lunar month, and earth's orbital year, and scripture could not be trusted for its astronomical accuracy.

But religious belief was not always so tolerant and could justify a much narrower approach to timekeeping. Denominational rivals could have their own preferred time systems, accusing others of corrupting God's true time. While Russell's study of heaven, both scriptural and astronomical, bent her towards flexibility, others used religious arguments to back more rigid beliefs about the proper way to keep time. Charles Piazzi Smyth, discussed later in this chapter, is one of these.

WILLIAM PARKER SNOW

While religious belief shaped some people's approach to time reform, others cited more practical concerns, especially among seagoing professions. Navigation and timekeeping had long been linked, as far back as 1714, when Britain's Longitude Act offered prize money for a method to establish longitude at sea. Determining latitude is fairly easy, but longitude is a challenge. One of the few methods available involved measuring the movements of the moon and comparing them to an almanac designed for the purpose, but this was clumsy and ultimately too finicky on a ship in constant motion. The only practical alternative was to compare the local time (at the ship's location) with the time at a known longitude on land. It's easy to determine local time, but not the time somewhere else on earth, except by carrying a clock. In the early 1700s, however, no clock could keep time accurately enough on rough seas over long voyages to be trusted.

Figure 2.2 | William Parker Snow (1817–1895), sporting his Arctic Medal.

It was clockmaker John Harrison who devised a clock accurate enough to do exactly that, allowing sailors to determine longitude at sea without the need for complex lunar observations. By the 1880s, Harrison's clocks had been improved upon many times over, and his method was used worldwide. But sailors wanted to further standardize timekeeping. William Parker Snow (1817–1895) (Figure 2.2), a lifelong sailor who took an interest in time reform in the 1880s, is a good example. Snow believed that standardizing timekeeping vis-à-vis a single prime meridian could improve safety at sea and prevent loss of life.

As his profession required, Snow had a working knowledge of navigation and practical astronomy. He led a troubled life, leading one scholar to conjecture some sort of mental illness.[47] Nonetheless, he was an intelligent author and experienced seaman. Recruited by Lady Jane Franklin in the early 1850s, he participated in several expeditions searching for the missing ships of her husband, Sir John Franklin, who had been lost in the Northwest Passage in 1845. The widely publicized search was a windfall for Snow, but it did not last. By the 1880s Snow had fallen on hard times, and was living in genteel poverty, supported by his publications and by relatives.

Snow had long hoped to find new ways to reduce deaths at sea. In November 1880, in *Chambers's Journal of Popular Literature, Science, and Art*, he proposed several lifesaving techniques.[48] He wanted to establish floating telegraph lines in the Atlantic, for example, to provide emergency communications. He also wanted unused naval ships to act as floating lighthouses, guiding ships into dangerous ports. Perhaps Snow's most ambitious plan was a series of "ocean relief depots" – small settlements or stores of food, water, and supplies placed in remote areas, to sustain stranded or lost crewmen until rescue arrived. He proposed the most desolate of places, like the St Paul's Rocks, in the middle of the Atlantic, or in particularly dangerous places like Cape of Good Hope (southern Africa), where many vessels were lost every year. Critics countered that such depots would be too expensive and difficult to maintain. The editors of *Chambers's Journal* found the depots "ingenious, but we fear not very practicable."[49] That's roughly what the Royal Society said about Fleming's time reforms: clever, but utopian. Like Fleming's ideas, Snow's proposals were not really so far-fetched, if only the political will had existed to carry them out.

When the subject of a common prime meridian for navigation and timekeeping rose to prominence in the early 1880s, Snow wanted to be involved. He immediately saw it as a possible lifesaving tool. He himself had experienced the confusion caused by its lack. During a storm at sea in 1832, the ship he was sailing came across a second vessel fighting its way through the gale. The two vessels traded their longitudes (this common practice helped each crew confirm its own calculations) and found that they were "wildly different."[50] Getting the longitude wrong in a storm with poor visibility was an often-deadly mistake. Indeed, the Longitude Act itself had been in part a response to a naval disaster near the Isles of Scilly, off southwestern Cornwall, in which four ships had sunk in a storm because they could not calculate their longitude. Snow implies that the two ships in that 1832 storm may have done their calculations correctly, but each using a different prime meridian. Snow was convinced that a shared international prime meridian would prevent such mistakes in the future.

To this end, Snow published a circular letter in 1883, addressed to the Royal Geographical Society and other interested parties, extolling the life-saving benefits of a common prime meridian. By contrast, he accused insurance companies like Lloyds of opposing a common prime meridian and the establishment of relief depots, which might cost insurance companies. If sea travel became safe, Snow insinuated, the firms would make no money, and so they lobbied against his plans.[51] Snow implored scientists to ignore these immoral corporations and support his ideas, which might save countless lives.

Snow's choice for the prime meridian was the St Paul's Rocks.[52] He thought it was ideal because it was neutral ground, avoiding any national jealousy. Furthermore, he wanted to turn these rocks into more than just an abstract line marked on maps and charts. The prime meridian could house one of his long-proposed relief depots. Responding to his critics at *Chambers's Journal*, he wrote: "As to its *practicability*, I argue nothing now. Man is the mighty spirit to surmount all difficulties; therefore I say, like Archimedes of Old – 'Give me a lever strong enough,' (money in this case,) 'and I could lift the whole world.'"[53] Here, Snow revealed his true purpose. He was looking for a patron to finance his writing. Snow concluded the circular: "How to *utilise* these Rocks, and form a PRIMARY *Station* there, as, also, one at the *Magnetic Poles* – North and South, – and other places, – I should be happy to explain, if means be afforded me, as I am too impoverished now to do it at my own expense."[54]

Snow's plans fell on deaf ears. While men faced fewer barriers to the sciences than women, engagement with professionals was difficult for any outsider. Professional masculinity had developed its own performative rituals and behaviour to regulate membership, marginalizing other masculine traits.[55] Professional societies were rife with ritual and symbolism, in an effort to legitimize their elitism, separating them from other men.[56] Expertise itself was not enough to guarantee acceptance. Without money or position or memberships, even a highly experienced person could be ignored.

Snow was unsuccessful. Despite his experience with navigation, and his first-hand understanding of the need for a common prime meridian, he gained little traction. He brought the language of nautical safety, a laudable cause, to the time-reform debate, but the Washington Conference would hear no mention of lifesaving at sea. Snow's exclusion was not due to a lack of expertise, nor due to class, race, or gender. It was a lack of position, and of funds. Just as Fleming's early papers were not legitimized until published by a professional society, Snow was an outsider looking in because he lacked official standing. He was an amateur.

Amateurism in Victorian science deserves some further examination. The gap between paid professionals and unpaid amateurs was growing wider by century's end, and the two sides had a conflicted relationship. Many fields were professionalizing, from the civil service to historians and lawyers.[57] In Britain, most astronomers were amateurs. Very few people actually earned a salary doing astronomy, apart from the staff of university and government observatories like Greenwich or Cambridge, the staff of the *Nautical Almanac*, and Norman Lockyer's research group in South Kensington, supported by the Science and Art Department.[58] For decades, amateurs had been the guiding lights of astronomy, and worked closely with professionals. But by late century, the two sides were drifting apart.

In some fields, like natural history, amateurs continued to find acceptance.[59] But in other fields, the gap widened. As historian John Lankford explains, "Professionals demanded specialization, technical knowledge gained only through advanced education, and access to large-scale research facilities. Further, they sought government support for their research, an action to which many amateurs strenuously objected."[60] Amateur astronomers, without access to major research observatories, argued that small, personal telescopes were more useful than the large, expensive variety used by their professional peers.

There was a sizable market for such cheap, portable instruments. In the 1880s, Josiah Latimer Clark, an electrical engineer and amateur astronomer, produced and sold relatively inexpensive (though still too dear for most working-class families) transit telescopes to the public.[61] Testimonials show a keen interest in both astronomical study and time-keeping. One customer from Pallasgrean, Ireland, wrote that operating it "would be quite a simple matter, and in this out-of-the-way part of the country will be most useful in regulating our time."[62] Accompanying the instrument was a yearly publication of transit tables, a simplified version

Figure 2.3 | Charles Piazzi Smyth.

of the official *Nautical Almanac*. The tables included instructions on the use of the transit instrument, and on how to determine accurate time, and to find Greenwich time, in any part of the world, and any day of the year.[63] With instruments like these available to the public, Greenwich did not have a monopoly on timekeeping, nor on astronomy in general.

Nonetheless, the amateur-professional divide was growing. Amateurs performed regular, steady observations, collecting data that made up the primary sources professionals used for their more advanced studies.[64] But professionals were seeking greater accuracy and technical skill, as well as recognition for their work above and beyond basic observation.

They invented new methods of separating themselves from amateurs, legitimizing their expertise with rituals, ceremonies, oaths, and organizations.[65] The end of the century marked not the end of the amateur, but merely the separation of amateurs from professionals. Relations between them could be friendly, but not always. The subject of this chapter's next section looked upon amateurs with disdain.

CHARLES PIAZZI SMYTH

Charles Piazzi Smyth, FRSE, FRS, FRAS, FRSSA (1819–1900) (Figure 2.3), was Scotland's astronomer royal from 1846 to 1888. He was born in Italy, and his middle name honoured his godfather, a Catholic priest and astronomer, Giuseppe Piazzi, who set up an observatory at Palermo. In his diary in November 1884, Smyth described a visitor to his Edinburgh Observatory on Calton Hill:

> At dusk, an unknown gentleman [a medical professor at the university] with a little boy 6 or 7 years old accosted me ... "When could I show my little boy your observatory, Calton Hill?" This is a good illustration of the extraordinary contempt and ignorance college people have of the nature of a Royal Observatory. I told him visitors were forbidden, that we were required to attend to our work; – and that he must get a telescope for himself ... But just as with ... Lord McLaren, wanting to make use of the Observatory through the winter, to save himself the expense of putting up a telescope in Edinburgh – so the wealthy medical professors at the college have no notion of spending their own money in getting themselves telescopes, even for their children.[66]

Smyth believed the professional's work was above public engagement and education. Although not all professionals agreed (Norman Lockyer regularly held public lectures, even for children), a clearly widening gap cordoned off both the public and the skilled amateur from the professional expert.

Smyth's antagonism to outsiders might stem in part from his own precarious position within the world of professional astronomy. He himself was in danger of being sidelined, even though he met all the criteria for membership. He was an affluent gentleman with both position and expertise. He was a competent astronomer, and a leading innovator in public timekeeping. His time-signal system in Edinburgh was one of the most advanced in the world, and foreign experts regularly went to him for advice.[67] It was his unorthodox opinions, not his talent or identity, that made him a target for exclusion.

As we saw in chapter 1, Smyth possessed a hypernationalist, religiously motivated obsession with the Great Pyramid of Giza, and his desire to see it become the world's prime meridian caused him to oppose Sandford Fleming's early proposals for standard time. Because he was Scotland's astronomer royal, his opinions, however outlandish, carried weight. We have seen how time reform was caught up in the relationship between science and religion through the works of Annie Russell, and how timekeeping was integral to navigation through William Parker Snow's campaign for nautical safety. An examination of Smyth's career adds several contemporary issues to the milieu. If Annie Russell represented a tolerant, respectable blend of science and religion, Smyth represented its fractious counterpart. His involvement also demonstrates how archaeology, metrology, imperialism, and international politics all intersected with time reform.

Smyth's pyramidology was a form of biblical literalism, at least insofar as he believed the accounts of the Hebrews in ancient Egypt accurate enough to compare them to the physical remains of the ancient past, such as the pyramids. He came across the subject when reading John Taylor's *The Great Pyramid; Why Was It Built: and Who Built It?* (1859), which argued that the edifice held instructions from God for the Anglo-Saxon race, and that the British system of weights and measures was divinely inspired. Reading Taylor's second book, *The Battle of the Standards* (1864), made him a convert.[68] This book framed pyramidology as a way to preserve Britain's national system of measurement in the face of the increasing popularity of the French metric system.

Smyth rapidly took up the cause, in 1864 publishing his own study, *Our Inheritance in the Great Pyramid*. Unlike Taylor, Smyth more carefully constructed his volume as a scientific argument, filled with diagrams, measurements, and citations. But he had not measured the pyramid itself. So in 1865, he decided to put his and Taylor's theories to the test, and travelled to Egypt with his wife, Jessica Duncan Smyth, to measure it for himself. Whether by coincidence or by way of self-fulfilling prophecy, Smyth's measurements confirmed for him that his theories were accurate.

What Smyth thought he had discovered was this: the ancient Egyptian cubit (a measure of length), which one would expect the architect of the pyramid to have used, when "applied either to the Great Pyramid's base-side, or base diagonals, or vertical height, or axis-lines, or any other known radical length of the building," does not come out to any round number.[69] However, if one uses a base of twenty-five British inches to measure the pyramid, it brings out "so many of the most important coincidences."[70] Given this striking discovery, Smyth concluded that the ancient architect must have used British inches; that somehow there was "intercommunication in idea and knowledge between the architect of the Great Pyramid, and the *origines* of the Anglo-Saxon race."[71] That architect was supposedly of Hebrew descent and had inherited the "sacred cubit" of twenty-five inches from Noah's son Shem. This argument drew a direct connection between the biblical Hebrews and modern Britons, thus imbuing the modern imperial system of measurements with both divine favour and ancient gravitas. The pyramid's purpose, according to Smyth, was to be a message to future generations, a time capsule showing the proper system of measure intended for man, as ordained by God.

Smyth's book added that the pyramid's measurements matched not just British length measures, but also measures of capacity, weight, temperature, and even time. He wrote, for example, that "when the British farmer measures his wheat which the bounty of Providence has afforded him as the increase of his land, in what terms does he measure it? In *quarters*. Quarters! Quarters of what? The existing British farmer does not know ... but from old custom, he calls his largest corn measure a quarter."[72] The modern British measure of capacity seemed to be exactly one-quarter of the contents of the coffer in the King's Chamber of the Great Pyramid. That unit of measure had somehow survived down four thousand years, and Smyth had now rediscovered its heavenly origin.

Similarly, he believed that the pyramid taught the correct use of the stars to tell time: "To astronomy ... only, of the modern sciences, can we reasonably look for some safe guidance in the practical measuring of time. In the broadest sense, time is said to be measured by the amount of movement of some body moving at an equable rate."[73] In the modern era, this was measured with a transit instrument, and Smyth claimed the pyramid contained its own ancient version. "To myself, who have been an astronomical transit observer for a great part of my life, it immediately occurred, that the narrow entrance-passage of the Great Pyramid directed up Northward, looked very much like a meridian *Polar* pointer."[74] Thus the divinely inspired ancient architect had applied astronomical methods "unknown elsewhere, and only recently begun to be appreciated in the best *European astronomy*."[75] Smyth's argument then becomes somewhat convoluted, suggesting that the pyramid's astronomical alignment may even be predictive, prophesying about events to come in the modern age. Today's Britons were to learn from and be guided by these discoveries, rejecting the metric system in favour of the imperial, and granting the Great Pyramid its rightful place as the prime meridian for global timekeeping.

The pyramid, in other words, was a form of religious scripture. "In an age when writing was a rarity," it offered a way to record sacred measures for posterity.[76] The Hebrew architect, inspired by God, foresaw a scientific age when careful study could unlock its secrets. Smyth himself spent 1865 conducting that study. Findings in hand, he returned to Scotland a true believer. It was his work, not Taylor's, that would come to popularize pyramidology. When time reform gained momentum in the early 1880s in preparation for the IMC, Smyth unsurprisingly put forward the Great Pyramid for the prime meridian.

Later historians and observers would scoff at Smyth's fantastical beliefs, labelling him "the Pyramidiot." But their dismissal of his pyramid research is somewhat ahistorical. Until the 1880s it was actually considered the most thorough ever undertaken and was praised by both professionals and the public.[77] Only later would it be debunked. For a time, pyramidology gained widespread acceptance, and its surprising popularity offers unique insights into Victorian culture writ large. It coincided with a broader fascination with Egypt, which was bound up in the imperial and international politics of the period, but also with archaeology, astronomy, religious belief, and time reform, all of which interacted in unexpected ways.

Egypt and its monuments held a prominent position in the public mind of Britain well before Smyth published his findings. If time reform captivated many Victorians, Egypt eclipsed it. "Egyptomania," fostered by Napoleon's campaigns in Egypt in the 1790s and the translation of the Rosetta Stone in the 1820s, spread from France to Britain and the United States after the Napoleonic Wars. The discovery of Tutankhamun's tomb in 1922 reignited it.

That fascination was alive and well in the 1870s and 1880s.[78] The British Museum and private collectors amassed (often illegally) large collections of Egyptian antiquities, and, with a steady flow of artefacts arriving in Britain, "Londoners of all classes ... could spend the afternoon gazing at curious foreign objects in public pavilions and exhibition venues."[79] Ancient Egyptian architectural styles began to find purchase in Britain, and stylized Egyptian motifs appeared on religious and commercial buildings, as well as in the monuments and architecture of cemeteries.[80] Large artefacts brought from Egypt became public monuments, like the inaccurately named "Cleopatra's Needle," an obelisk moved from Alexandria to the bank of the Thames in 1878. The Needle, in spite of its own ancient inscriptions, was dedicated instead to the memory of

British military victories in Egypt. The monument's erection in London, like Smyth's pyramidology, tied ancient Egyptian prowess to modern British nationalism.

A seemingly endless stream of new discoveries spurred this obsession with Egypt. Archaeological finds brought media attention, and artefacts that seemed to confirm biblical events, such as the Pithom stele, drew especially high levels of public interest.[81] Just as Annie Russell looked to astronomy to explain biblical passages, others turned to archaeology to do the same. As historian David Gange points out, "In the 1880s a biblically inspired constituency provided the bulk of Egyptology's readership: they revelled in the proofs and illustrations of the bible that archaeologists, year on year, appeared to unearth."[82] Smyth's pyramid measurements carried a similarly religious appeal: they seemed to confirm or supplement interpretations of the Old Testament.

British "orientalist" imaginings of the ancient Egyptians changed over time.[83] In the mid-nineteenth century, many British Christians reviled them and their monuments, as they represented a cruel, pagan civilization forsaken by God. As one of the era's Egyptologists put it, the later monuments were built by an "idolatrous monarchy."[84] Smyth himself held this view. He carefully distinguished the Great Pyramid, supposedly a sacred structure built by God's chosen people, from other monuments. Smyth remarked that "the Great Pyramid, though *in* Egypt, is not, and never was, *of* Egypt."[85] Other monuments in that country were not worthy of the same reverence. The Old Testament's portrayal of its rulers as the nemesis of God's chosen people, and their worship of an enigmatic pantheon of unfamiliar gods, did little to endear their civilization to British Christians. But their exotic "Otherness" provoked as much interest as disdain. The imagined Egypt of idolatrous tyrants captivated some members of the public. Smyth's vision of a sanctified and holy pyramid was doubly attractive, exploiting this fascination with the "Other," while removing the negative connotations of idolatry. It was an enchanting combination.

In the 1880s, Victorian perceptions of Egyptian civilization became much more favourable.[86] This shift has not only cultural, but also political explanations. Promoting the ancient land's history, archaeology, and the preservation of its monuments suddenly became politically expedient in 1882, when Britain needed to justify its occupation of the country. Its significance to British foreign policy in the 1880s was enormous. Even before the occupation, it was expected to play a role in resolving the "Eastern Question," the perceived weakening of the vast Ottoman Empire. Egypt was independently ruled, but under nominal Ottoman control. Its stability was considered an invaluable counterbalance to the ongoing uprisings and conflicts in the Ottoman Balkans. Britons had long travelled through Egypt on their way to India, but the opening of the Suez Canal in 1869 provided a shorter, easier route.[87]

British rule made Egyptologists and pyramidologists alike overjoyed. Smyth wrote that the opportunity to study the pyramid alone justified the British occupation, and others shared his enthusiasm.[88] One admirer wrote to Smyth: "Is this not wonderful news about our purchasing Egypt? – Fancy the Pyramid ours!! – I should like to go there if this is to be."[89] Another wanted to examine it more carefully. "If our intervention in Egypt succeeds and we reach Cairo as conquerors, the scientific members of the force would have an opportunity of remeasuring the four sides and setting the question at rest once and for all."[90] One woman told Smyth's wife, Jessica Duncan, that the pyramid study was as important as the "great Eastern Question" itself.[91] The idea of a British Egypt stirred enormous excitement. The general public was entranced by the intervention there and the subsequent Sudan crisis. The exploits of General Charles Gordon and his doomed last stand at Khartoum inspired poetry and art, and visceral fascination.[92] They made others angry. Astronomer J.C. Adams, the staunch conservative who represented Britain at the IMC, was particularly incensed at William Ewart Gladstone's Liberal government for failing to follow up early successes.[93] Either way, Egypt was at the heart of British foreign policy in the early 1880s, and everyone knew it.

Its centrality to Victorian Britain helps explain how a far-fetched theory like pyramidology could entrance so many people. Waynman Dixon, one of Smyth's more upstanding devotees, was an engineer and amateur Egyptologist who designed the vessel that carried Cleopatra's Needle from Alexandria to London. In early 1877, he wrote to Smyth:

Though I admit that I consider [Cleopatra's Needle] of great historic interest, and think a good thing was done in removing it to preserve it from destruction and loss, I candidly confess that £10,000 might have been very much better spent, especially on the further investigation and opening up of the Gt. Pyd. – Will no one come forward with a like sum for this purpose? ... I cannot help believing we are on the eve of some great political crisis arising out of the Eastern Question, in which Egypt will play a prominent part, and I should not be surprised if what I have long foreseen – namely an English occupation or protectorate of that country should be one of the first events to be noted. It would be a glorious deliverance from bondage

to the Egyptians and a commencement of the blessings foretold for
that oppressed country. And if such be the case will there not be a
great work before the Gov't to perform in the clearing up of the
doubts – to my mind – still hanging over the Great Pyramids origin
and meaning?[94]

Dixon concluded by asking for any new updates on pyramid research:
"How does the transit of Venus accord with Pyramid measures?"[95]
The letter makes clear Egypt's crucial place in British foreign policy
and culture, predicts the coming crisis with remarkable foresight, and
dreams of opportunities for further archaeology. It also depicts pyramidology as a scientific endeavour, vis-à-vis the transit of Venus. Believers
treated the pyramid as a legitimate source of information about the
universe and its workings, including how to measure time and space.
Seen in this light, Smyth's pyramidology was a tested scientific hypothesis, which tapped into both public cultural fascinations and political
machinations, as well as religious and nationalist sentiment, thus gar-
nering widespread acceptance. It was an attractive, compelling idea,
so when Smyth suggested that the pyramid would make a good prime
meridian, people listened.

To get some sense of the popularity of pyramidology, take one remarkable statistic uncovered by historian Eric Reisenauer. Between 1883
and 1886, in Barrow-in-Furness, in Cumbria (northwest England), 147
people checked out Smyth's book, *Our Inheritance in the Great Pyramid*,
from the Free Public Library, which averages out to about one borrower
a week. The majority of the readers, apparently, were from the working
class.[96] Smyth received plenty of correspondence from readers, asking
for permission to give public lectures on his work, or to convert his
book into Sunday school lessons, or print cheap pamphlets that would
summarize his book for the masses.[97]

But his theories found the most purchase with a few particular nationalist and religious sub-cultures.[98] The first was the anti-metric movement,
made up of men like Charles Latimer (leader of the IPAWM in the United
States), who joyfully accepted Smyth's work as a new justification for its
cause. What greater argument for imperial units of measure than their
being ordained by God? Latimer told Smyth, "You have given us the
weapon to utterly overthrow the French metric people ... We will work
until we overthrow the French meter in its own house."[99] In 1864 and
1878, the British Parliament debated reforms to the measurement system, and subsequent bills even legalized use of the metric system. These
events gave the anti-metric movement a sense of urgency, and Smyth's

theories were a ready-made counter-argument.[100] As Latimer put it, "The only weapon that can ever prevail over the French meter is the scriptures and the Great Pyramid."[101]

Meanwhile, another nationalist, but more religiously motivated movement took up pyramidology with equal vigour: the British Israelites.[102] Adherents believed that the Anglo-Saxon peoples were the direct descendants of the tribes of Israel, and therefore God's chosen people. Their circular reasoning both justified British imperial expansion as God's will and made it a validation of their belief in British superiority.[103] Their rise in popularity coincided, not surprisingly, with a renewed imperial identity in Britain.[104] Smyth's theory that the British had inherited divinely inspired measurements directly from Israel, as demonstrated by their appearance in the ancient pyramid, was all too easily adopted into the teachings of the British Israelites.

Pyramidology found its way into unexpected places in both British and American culture. Time reform was not immune. Influenced by Smyth's theories, the IPAWM joined the U.S. time-reform movement, in an attempt to secure the Great Pyramid's place as the prime meridian. Charles Latimer adopted Sandford Fleming into the IPAWM and used that connection to keep tabs on his time-reform activities. Through its contacts on both sides of the Atlantic, the IPAWM attempted to ensure that any decision-making regarding a prime meridian would at least consider the pyramid (and its divinely derived British inch). In the years leading up to the IMC, the IPAWM lobbied the U.S. government and conference organizers, hoping to fit the pyramid into the emerging system of global timekeeping. And the IPAWM was not alone. Letters from believers flooded into the IMC organizers from all over, lobbying for the prime meridian to be placed at Giza.

As preposterous as pyramidology may have been, its followers showed considerable ability to influence time reform. Its sizable public following was matched by support within the scientific community, at least for a time. In Victorian science, religious or divine phenomena were tested by the scientific method just as any other natural phenomenon would be. The *London Quarterly Review*, for example, showed a healthy agnosticism towards Smyth's first publication on pyramidology, deciding to withhold judgment until after he had finished his measurements in Egypt.[105] With his results in hand, he seemed to have a compelling case for his theory and, for a decade or so, convinced at least some scientists. Upon his return from his 1865 trip to Egypt, he presented his findings to the Royal Society of Edinburgh, and he was awarded the Keith Medal, a biannual prize for the most significant paper published in the Society's journals.[106]

Eventually, Smyth did fall out of favour. Further surveys of the pyramid contradicted his findings. In 1874, Scottish-born astronomer David Gill, surveying in Egypt after being in Mauritius to observe the transit of Venus, measured parts of the pyramid in his spare time, but could not complete the work, although he and colleagues placed a wooden mast on top of the structure, apparently to approximate its original height before the marble stones eroded. Gill's incomplete survey sparked a second expedition in 1880, led by Flinders Petrie, who hoped to confirm Smyth's findings. Petrie's father had been Smyth's friend, and both Petries believed Smyth to be correct.[107] But Flinders Petrie discovered that the regular Egyptian cubit could explain the pyramid's dimensions, pointing out several mistakes by Smyth that undermined the sacred cubit of twenty-five inches.[108]

Petrie's measurements, the pro-metric lobbying of F.A.P. Barnard in the United States, and the criticism of another prominent astronomer, Richard Proctor, ruined Smyth's reputation. Smyth could have accepted his mistake and changed his hypothesis to fit the new evidence. Instead, he doubled down on pyramidology, turning a blind eye to critical peers. This cost him the respect of mainstream science and forced him to embed himself instead into the hypernationalist theological discourse of British Israelism and anti-metric fanaticism.[109] It was his abandonment of scientific principles for unfettered mysticism that alienated him from the astronomical community, not pyramidology itself. Had the IMC taken place fifteen years earlier, before new evidence came to light, Smyth's case for the pyramid as prime meridian might have been taken more seriously.

As it was, the astronomical community's rejection of pyramidology was a turning point for Smyth. His work was attacked in the Royal Society's *Proceedings* by Sir Henry James, who had once called Smyth's theory "sheer nonsense in a comically solemn dress."[110] Smyth wanted to respond, but the Society's secretary wrote that Smyth's rebuttal was "not of a nature suited for a public reading before the society."[111] Insulted by the snub, Smyth wrote an indignant letter of resignation.[112] It was the first time any member had ever resigned.[113] As historians Mary Brück and H.A. Brück suggest, it brought him ridicule and, more important, "put a barrier between him and the influential scientists of the 'establishment.'"[114] Smyth retreated to other less reputable, though friendlier, forums, such as Latimer's IPAWM.

Smyth did not cut off all personal relations with Royal Society members. Networks of science were personal as much as professional. Astronomer J.C. Adams, for example, continued to dine with Smyth on occasion, and regretted that he had resigned. James R. Napier wrote to

Adams in early 1876: "I have not seen our friend Piazzi Smyth I think since I dined with you and Mrs. Adams at his house. I wish he had not left the Royal Society in the way he did – I wish he would let the Pyramids alone – for I dare say you and your telescopes with a little friendly intercourse with Venus can tell us more about the Sun's distance than that Egyptian monument he has made so much of."[115] Napier and Adam's contacts with Smyth helped ease the break, but Smyth's ideas were left behind in the face of more promising scientific endeavours like the transit of Venus.

It took more than a decade for Smyth to be entirely ousted from his position. He was still a competent astronomer, and a leading light on the subject of civic time signals, and he was regularly consulted on time distribution. But in his last few years as astronomer royal for Scotland, he found himself increasingly isolated. His final notes as astronomer royal in his Equatorial Book from 1888 are morose, though also defiant, including stories about being pelted with stones by "mischievous boys" on his way home in the evenings, complaints about being taunted by the other professors of Edinburgh University, and advice to his successor to be watchful about anything "unexpected and untoward."[116]

Before his exit, however, Smyth's pyramid evangelism helped the astronomical community reject Fleming's time-reform scheme. His tenuous position as astronomer royal allowed him, alongside George Airy, to convince British astronomers to dismiss Fleming's time proposals in 1879. Indeed, by tabling the subject when he did, Fleming gave Smyth the opportunity to use his platform to champion the pyramid as prime meridian. For the next five years, the pyramid regularly surfaced in legitimate forums as a viable prospective location for the meridian. This was not the work of Smyth alone. Egypt's hold on the consciousness of Victorian Britain infected the time debates, right up to and including the Washington Conference. While the pyramid was ultimately a dead end, it lay at the confluence of political, religious, social, and scientific debates – a perfect reflection of the cultural milieu in which the time-reform debates at the IMC were embedded.

CONCLUSIONS

Time reform did not occur in a vacuum. It was influenced by a variety of competing interests. Most professional astronomers, for example, saw time standardization as a highly specialized task. Their experience with observations of the transit of Venus in 1874 and 1882 taught them the value of careful, precise timekeeping, and made them understand the need for a global time system. To them, it was a tool that facilitated complex observations, or for navigation at sea, but had little use outside their profession. At Washington in 1884, they found themselves entirely at odds with reformers like Fleming, who wanted to change the way everyone kept time. Asking the general public to adopt new norms of time measurement in their everyday life seemed to them unnecessary and impossible.

That duality – universal time as a scientific tool and/or for everyday life – formed the core of the time-reform debates. But the cases of Russell, Snow, and Smyth reveal a richer and more complex tapestry of ideas surrounding that core. Archaeology, metrology, and theology shaped perceptions of timekeeping as much as astronomy did. In a similar vein, areas of expertise as varied as nautical safety and scriptural interpretation offered legitimate gateways into the controversy. These various discussions about time among professionals and amateurs alike were also transnational, easily crossing national borders. Of course, colonialism and prejudice affected everything. In professional astronomy, for example, when British astronomers asked the opinions of the astronomical community in Cape Colony, they were not really seeking out local expertise, but rather engaging with David Gill, a Scottish-born scientist who served 1879–1906 as her/his majesty's astronomer at Cape Town. Similarly, social class limited participation in the sciences, perhaps even more than race or gender. Wealthy women and colonial subjects, though facing various forms of discrimination, could sometimes make a space for themselves in the sciences if they wished. Working-class people usually could not. Even amateur astronomy, for example, was prohibitively expensive. This made the astronomical community quite small and insular.

But the margins of the astronomical network were vibrant and stimulating places, populated with both skilled and eccentric personalities, including Russell, Snow, and Smyth, as capable of producing knowledge as "insiders." However, as they were excluded by social norms and regulations, both formal and informal, finding a platform for their work was hard. The knowledge they produced was often not considered "authoritative" so carried less weight in policy decisions, but nonetheless broadened the time debates in interesting ways. These outsiders' experiences show how various topics shaped timekeeping, ranging from navigation and astronomy to religion and archaeology. Religious belief could either offer a sense of lighthearted apathy about human timekeeping, as in the case of Russell, or drive a person to extreme, unyielding opinions, as with Smyth. Occupational expertise, such as Snow's seamanship and concern for safety, offered unique perspectives on time reform. All of this occurred in a culture searching for meaning from both modern science and ancient texts, looking to understand the world through a variety of means. This swirling sea of beliefs and ideas came to a head in 1884 at the IMC, where time reform could not free itself of its baggage. In particular, the IMC's results reflected the struggle between amateur and professional (as well as among professions), as insular networks tried to drown out less influential ones. The rise of professionalization often left outsiders' ideas about time unheeded. Yet they still had some influence, filtering into the IMC discussions in small ways. Their ideas enriched the otherwise-simplistic competition between the railway engineers' vision of universal time for civil life and the astronomers' vision of universal time for scientific endeavours. How those differences in aim and scope played out in the context of international diplomacy at Washington in 1884 is the subject of the next chapter.

3

The International Meridian Conference

In 1884, Washington, DC, was a city in transition. It had emerged from the American Civil War in 1865 with dirt roads and poor infrastructure. Characterized by poverty and crime, it was a far cry from the grandiose capital that its planner, Pierre Charles L'Enfant, had envisioned.[1] Built on a wetland that occasionally flooded, and plagued by periodic political instability, it was – and this was the most complimentary thing one could say about it a century after its founding – still a work in progress.

However, there were signs of renewed growth, although growth came at the price of deep debt. Governor Alexander Robey Shepherd (in office 1873–74) began a series of massive public works, including new paved roads, streetcars, public buildings, and monuments. The Washington Monument, which had long stood unfinished in the midst of civil war, was finally completed at the end of 1884. As the country's most ostentatious appropriation of Egyptian architecture, the obelisk was at the time of completion the world's tallest structure, combining nationalist airs of inherited civilization with modernist messages of progress and power.

Another of the gleaming symbols of the "new" Washington was the State, War and Navy Building (Figure 3.1; now known as the Eisenhower Executive Office Building), across from the White House, built 1871–88 in the French Second Empire style and for years the world's largest office building (with 566 rooms). In 1884, it too was a work in progress. The south wing, which housed the Department of State, was the first side finished, and had been in use since 1875. More recently, the Navy Department had moved into the east wing. The rest of the building was still a construction site. The south wing housed a great, gilded Diplomatic Reception Room, which was designed to be the nation's showpiece to foreign diplomats. In this grand hall, the swampy roads

Figure 3.1 | Executive Office Building in Washington, DC, site of the IMC.

outside could be forgotten, and the business of states could be trans-
acted. In October 1884, that business was the establishment of a shared
global prime meridian.

The International Meridian Conference (IMC) that took place in
Washington that year has a mixed reputation. Historians sometimes dismiss it as an event "largely without impact," and they are not entirely
wrong to do so.[2] The notion that standard time was established at the
IMC is fanciful at worst, and an oversimplification at best. Global timekeeping was ultimately organized piecemeal by individual nations over
a period of fifty-odd years – international efforts like the IMC did little
to shape those national policies. But why did such international efforts
fail? Twenty-five nations came to the table in 1884 to discuss a prime
meridian and its uses, only to leave without a consensus on timekeeping.
What happened?

Historians tend to place the blame on friction between France and
Britain.[3] These countries were certainly the two most vocal opponents at
Washington, having come to the conference with incompatible proposals

for the location of the new prime meridian. But a more careful reading of conference events reveals that very few delegates, regardless of nationality, even wanted standard time.[4] The few who did so, such as Sandford Fleming and William Allen, were in the minority. These two representatives of North America's commercial and engineering interests were not able to overcome the misgivings of the astronomers, navigators, and diplomats who were interested mainly in the nautical and astronomical applications of a prime meridian, rather than its role in civil timekeeping.

This conflict between engineers and astronomers has been missing from historical discussions of the IMC mainly because it is not obvious – it is one thread among dozens at play. The IMC was a moment when layers of conflicting interests clashed: timekeeping v. navigation, astronomer v. engineer, amateur v. professional, French v. British, science v. religion, Christianity v. Islam, metric v. imperial, and so on. These rivalries, which shaped and informed the era's time-reform debates writ large, remained relevant at the microcosmic scale of the conference. The multitude of debates simmering beneath the surface in contemporary culture came to a head at Washington, making for a messy event.

The task of unravelling these threads is daunting. A good starting point is the conference's *Protocols of the Proceedings* (1884), a two-hundred-page document that records many speeches made by delegates, the resolutions they passed, and which nations voted for them. Its comprehensive, though not always verbatim account tells us what happened, but not necessarily why. If we want to know why, then context is key – chapters 1 and 2 have helped us identify some of the threads, allowing us to pick out the most significant. We have seen what motivated several of the participants. In some cases, such as Fleming's, the IMC was the climax of six years of campaigning for a change to civil timekeeping. A few others, such as J.C. Adams, arrived at the conference with little knowledge of the timekeeping campaign, having been chosen by Britain's Science and Art Department solely because of their anti-metric outlook. Others still were diplomats, non-specialists there to represent their nation's interests, without any knowledge of the particulars of the subject. Lionel Sackville-West, for example, the British ambassador to the United States, was in charge of greeting and coordinating his country's delegation, but knew little about the subject matter. The brother of delegate Sir Richard Strachey, for Britain (India), warned him: "West is a very good fellow, but would hardly understand 'geodetic' without looking it up."[5]

The net result was a massive imbalance between participants in terms of their investment in the subject matter. Very few delegates went to the IMC to talk about timekeeping, and this explains why Fleming's conflict with the astronomers has been largely forgotten. The battle between France and Britain, which had its roots in the metric debate, attracts the most attention, because it seems to be the biggest conflict at the conference. But to answer our central question – why did global standard timekeeping take the form that it did? – we need to shift away from that Anglo-French conflict, and focus instead on the attempts to introduce civil timekeeping into a debate that was otherwise focused on navigation. Similarly, we must examine some of the more obscure conference resolutions, which relate to changes in the "astronomical day." The debates over that subject, both then and later, reveal again that astronomers saw accurate, standardized timekeeping as a specialized tool, unrelated to civil timekeeping.

In order to obtain access to these undercurrents, this chapter draws from personal papers in addition to the conference proceedings, so we can examine the "culture" of the event. What was going on during the days between the official meetings? Who talked with whom, and where? The IMC was a gathering of the international astronomical community with all its preconceptions and traditions and rules of membership, but complicated by the presence of diplomats, engineers, and lawmakers in its midst. By focusing on the personal and relational experiences of the decision-makers, we can begin to grasp the motivations for their decisions, apart from any abstract notions of "national interest." The drawing together of all these threads makes for a very different picture of the IMC than a simple French-British rivalry might suggest. Profession, not nationality, decided the outcome of the conference.

THE IMC BY THE NUMBERS

The IMC lasted exactly one month, from 1 October to 1 November 1884. During that time, eight sessions were held, of which the first and last were mainly ceremonial and organizational. The real work was done during the middle six sessions. Forty-one delegates from twenty-five countries attended. A twenty-sixth nation, Denmark, was expected to participate, but its consul-general, Carl Steen Anderson de Bille, never showed up. Of those countries that did participate, eleven were European, ten were South American or island nations, two were North American, one was Asian, one was African, and two "colonies" (Canada and India) were present under the banner of the British Empire – these were all governments with which the United States had diplomatic relations. The United States' long-held Monroe Doctrine stated that the western hemisphere ought to be its exclusive sphere of influence and partially explains

Table 3.1 | Occupational groups of IMC delegates

Occupation	Number Present	% of total
Diplomats	20	48.7
Navy	8	19.5
Scientists	7	17.1
Engineers	4	9.8
Surveyors	2	4.9

Note: This breakdown by occupation is somewhat simplified, because these categories occasionally overlap: some naval representatives were also astronomers, and so on. S.R. Franklin and Richard Strachey are two examples.

Table 3.2 | Countries with an IMC delegate who spoke more than five times

Country	Delegates
Britain	Adams, Evans, Fleming, Strachey
France	Janssen, LeFaivre
Russia	Struve
Spain	Arbol, Pastorin, Valera
Sweden	Lewenhaupt
USA	Abbe, Rodgers, Rutherfurd, Sampson

the heavy participation of South American nations. But their delegates arrived in Washington with their own ambitions and alliances, so not all voted in line with the United States. Some, like Brazil and San Domingo (later the Dominican Republic), often sided instead with France.

Some countries were less committed to the process than others. A few missed parts of the conference: the Turkish delegate did not arrive until the third session; Chile, Holland, and Liberia's delegates arrived in time for the fourth; and Salvador missed the fourth, sixth, and seventh. There were also significant variations in the number of delegates each country

Table 3.3 | Number of times each IMC delegate spoke

Name/country	Occupation	Times delegate spoke
Schaeffer/Austria	Diplomat	3
Cruls/Brazil	Scientist	2
Adams/Britain	Scientist	19
Evans/Britain	Navy	6
Fleming/Britain (Canada)	Engineer	9
Strachey/Britain (India)	Diplomat (among other skills)	22
Gormas/Chile	Navy	0
Tupper/Chile	Navy	1
Franklin/Colombia	Navy	1
Echeverria/Costa Rica	Engineer	0
Bille/Denmark	Diplomat	Did not attend
Janssen/France	Scientist	21
LeFaivre/France	Diplomat	19
Alvensleben/Germany	Diplomat	4
Hinckeldeyn/Germany	Diplomat	0
Rock/Guatemala	Surveyor	1
Aholo/Hawaii	Diplomat	0
Alexander/Hawaii	Surveyor	0
Foresta/Italy	Diplomat	1
Kikuchi/Japan	Scientist	0
Coppinger/Liberia	Diplomat	0
Anguiano/Mexico	Scientist	0

Table 3.3 | Number of times each IMC delegate spoke (*cont.*)

Name/country	Occupation	Times delegate spoke
Fernandez/Mexico	Engineer	1
Weckherlin/Netherlands	Diplomat	0
Stewart/Portugal	Diplomat	1
de Struve/Russia	Diplomat	9
Kologrivoff/Russia	Diplomat	0
Stebnitzki/Russia	Diplomat	0
Galvan/San Domingo	Diplomat	1
Batres/Salvador	Diplomat	1
Arbol/Spain	Navy	8
Pastorin/Spain	Navy	5
Valera/Spain	Diplomat	14
Lewenhaupt/Sweden	Diplomat	15
Frey/Switzerland	Diplomat	2
Hirsch/Switzerland	Scientist	Unclear if he attended
Rustem/Turkey	Diplomat	4
Abbe/USA	Scientist	8
Allen/USA	Engineer	3
Rodgers/USA	Navy	107 (president/moderator)
Rutherfurd/USA	Scientist	29
Sampson/USA	Navy	15
Soteldo/Venezuela	Diplomat	2

Table 3.4 | Number of times IMC delegates spoke, grouped by occupation

Occupation	Times delegates spoke	% of total speeches
Diplomats	98	43.2
Scientists	79	34.8
Navy	36 (excluding Pres. Rodgers)	15.9
Engineers	13	5.7
Surveyors	1	0.4

decided to send. The cost of travel led some small nations to send fewer delegates, most relying on their permanent ambassadors to Washington. Another important factor at play, however, was the last-minute U.S. concession to Britain allowing it five delegates instead of three. This change, made mere months before the conference began, left little time to make plans for new delegates. In the end, of the twenty-five countries that attended, only nine sent more than one delegate, most of which were European, and of these only Britain and the United States exceeded three.

The *Protocols of the Proceedings* allows us to analyze delegates and their activities at the conference. Delegates came from five general areas of expertise: twenty were diplomats, eight had naval backgrounds, seven were scientists, four were engineers, and two were land surveyors (see Table 3.1).[6] Five non-voting experts were invited to speak as well, all but one of whom were scientists.[7]

It was the scientists who spearheaded the discussion. The proceedings allow us to assess the number of times each delegate spoke. "Number of times spoken" is not a perfect variable: a lengthy speech receives the same weight as someone suggesting a break in the proceedings. The numbers are also skewed by the fact that Admiral C.R.P. Rodgers, the American delegate, spoke 107 times as president of the conference, but he was merely moderating the discussion. Nonetheless, a count like this is a useful way to gauge who was most involved. For example, the only countries with delegates who spoke more than five times (Table 3.2) were

Britain, France, Russia, Spain, Sweden, and the United States, implying that this was a heavily Eurocentric gathering.

If we break number of addresses down by occupation, diplomats spoke ninety-eight times, scientists (not including the non-voting experts) seventy-nine, navy representatives thirty-six (excluding President Rodgers), engineers thirteen, and surveyors once (Table 3.3). Although scientists made up 17 per cent of delegates present, they spoke 35 per cent of the time – the only occupation to punch above its weight at the conference in terms of verbosity (Table 3.4). In other words, diplomats spoke more times than scientists, but only because there were more of them. Statistically, the most vocal delegates by far were the scientists.

BEYOND THE NUMBERS

Statistics take us only so far. Historical context provides further insights into the makeup of the gathering. There were dozens of nations, kingdoms, and other types of polities in existence in 1884 that were not invited. Many of these were not recognized as sovereign entities in European or American eyes. Indeed, quite the opposite, as is evident from the other diplomatic conference of significance to take place in the autumn of 1884: the Berlin Conference. Beginning in November, just days after the IMC ended, that conclave saw the European powers carve out arbitrary spheres of influence for themselves in Africa. This formalized "scramble for Africa" demonstrates the lack of recognition Europeans gave to African polities and explains their absence from the IMC. The one exception, Liberia, was in a way both an American colony and a colonizing power itself, albeit in a strange way. Liberia was created by Americans in an effort to "repatriate" freed slaves to western Africa in the 1840s. Liberia kept close ties with the United States even after its independence, guaranteeing its place in U.S. foreign policy, and earning it a certain amount of patronage. Indigenous African polities found no such favouritism, let alone recognition of their sovereignty.

Because of the two events' temporal proximity, several scholars have drawn connections between them.[8] While Berlin set the terms for western exploitation, Washington enabled it.[9] Implementing standard time temporally reinforced a core-periphery relationship, with London at the centre of both the temporal and the material world.[10] "The project to incorporate the globe within a matrix of hours, minutes and seconds," writes historian Giordano Nanni, "demands recognition as one of the most significant manifestations of Europe's universalising will … The conquests of space and time are intimately connected."[11] Historian

Vanessa Ogle echoes this sentiment, cautioning that globalizing processes, such as standardizing time, are not neutral but ideological: "When Europeans and Americans wrote about time-and-space-defying connections and uniform time as a means to bring order to a globalizing world, they proposed to create a world in their own image and a world of their own domination."[12] This world was hierarchical, rather than equitable.

The geographical makeup of the IMC delegates bears out these insights. Smaller South American nations, though better represented than African polities, were similarly vulnerable to the whims of their more powerful neighbours, particularly the United States. For example, the assertion that Colombia sent a delegate to the IMC is somewhat deceptive. It was represented by an American naval astronomer, Commodore S.R. Franklin, superintendent of the U.S. Naval Observatory in Washington, DC, since February 1884. In February 1883, the director of the Astronomical Observatory of Bogotá wrote to the U.S. Naval Observatory requesting that the latter's superintendent choose someone to represent Colombia at the conference. He assumed that, "on account of the aproximity [sic] of the two meridians of Bogota and Washington, [the interests of Colombia] must be naturally identical with those of the United States of America."[13] The U.S. Naval Observatory did nothing to disabuse the Colombians of this assumption, despite the fact that since the request had been received, the United States had replaced the meridian of Washington with Greenwich. Colombia was not notified who was representing it until 14 October, after the IMC had begun.[14] During the conference, Franklin barely spoke, and he voted in line with the United States on every major resolution. In his report to the Colombian government, he was matter of fact, stating the results of the IMC without offering any explanation for voting the way he did.[15]

Similarly, Guatemala was represented by American-born surveyor Miles Rock. As president of the Boundary Commission that established the Mexico-Guatemala border in 1883, Rock had helped Guatemala retain some of its contested frontier territory, despite Mexican ambitions. That work earned him the respect of its government and explains his appointment. Another similar case, though not South American, is that of Hawaii. One of its representatives was a Yale graduate, and a member of the island kingdom's American and European elite. Soon after the IMC, a coup instigated by this elite class overthrew the monarchy, and the islands were annexed by the United States before the century's end. The cases of Colombia, Guatemala, and Hawaii, though not overtly malicious, show how larger nations could appropriate or undermine the representation of smaller nations at the IMC.

What does this mean for our analysis? It shapes the lens we use to examine the conference. Scientific endeavours are never politically neutral. The choice of a prime meridian was no exception. The IMC and the Berlin Conference are more alike than they are different, both being integral to the western project of globalization, with all its colonialist underpinnings. It is in this context that we must place the actors who arrived at Washington in the autumn of 1884.

ARRIVALS

For many IMC delegates, the expedition to Washington lasted several months, in which the conference was just one stop of several. Most had other professional obligations as well. For example, Kikuchi Dairoku's expedition to the United States in 1884 was as much about scientific networking as diplomacy. Kikuchi (the Japanese delegate to the IMC) was a professor of mathematics at Tokyo Imperial University and had studied at Cambridge. His trip to the United States allowed him to attend physicist Lord Kelvin's master-class lectures on molecular dynamics at John Hopkins University in Baltimore. The prime-meridian question occupied only a small portion of Kikuchi's time in the United States.

Kikuchi's experience was not unique. Many other delegates made similar calls on overseas colleagues and participated in other professional gatherings. The British Association for the Advancement of Science (BAAS) held its annual meeting in Montreal in August and September 1884, the first time it had ever been held outside the British Isles. British delegate Professor J.C. Adams attended it, as did Sandford Fleming, this time as vice-president of Section G, "Mechanical Sciences."[16] Sir Richard Strachey considered going as well.[17]

After arriving in North America early for the BAAS gathering, Adams and his wife went on a whirlwind tour, combining both pleasure and academic outings. They visited wilderness destinations in Quebec and then gazed in awe at Niagara Falls, which Adams recalled was "grand beyond description."[18] They then attended a scientific meeting in Philadelphia, where they stayed at a ladies college, having been invited by a "lady professor of mathematics," Mrs Cunningham, whom they had met previously in Cambridge.[19] The pair travelled through the Alleghany Mountains and took a cruise on Lake Champlain before heading at last to Washington. Adams wrote to his friends back home, saying that, although he enjoyed the trip, he was disgusted by the newspaper coverage of the ongoing American presidential campaign (voting was set for 4 November), which was full of "the vilest personal attacks."[20]

Sandford Fleming also travelled with his wife, Jeannie. They arrived in Washington via New York on 30 September.[21] While her husband was in conference sessions, Jeannie took an excursion to Mount Vernon. Once the home of George Washington, it was a popular tourist destination for affluent visitors to Washington. Fleming had already been there – in May 1882, after a meeting with members of Congress about organizing the IMC, he had boarded a cruise ship and travelled down the Potomac to Mount Vernon. At that time, he had written that he wished Jeannie was there to go too.[22] Now was her chance.

Tourism and networking went hand in hand. French delegate Jules Janssen made his own networking visits upon arrival in the United States. He gave interviews with New York reporters (lamenting that he did not have his wife as interpreter, as he spoke only French), and dined with telephone inventor Alexander Graham Bell. He also met with Frederick Barnard, the president of the American Metrological Society, who, as we saw above, would have been an IMC delegate but had resigned due to hearing loss.[23] These visits mixed leisure with professional social calls, and very few of either had anything to do with timekeeping and the prime meridian.

The above itineraries of several delegates should make it clear that the IMC did not take place in a vacuum. The conference was embedded in the era's scientific network, which was in the process of forming and consolidating professional organizations through personal relationships. The IMC was therefore wrapped up in the trappings of professional science, as much if not more than it was in diplomatic and political procedures. It is significant that the conference lasted thirty-two days, but held formal sessions on only eight. Delegates would have spent much of the time in between preparing, translating, and reading proposals for discussion, but also socializing and networking.

Many of the IMC delegates (see group photograph in Figure 3.2) were diplomats similarly ensconced in their own networks and negotiations. Manuel de Jesús Galvan, the envoy from San Domingo, for example, was negotiating a trade deal between his country and the United States. On days between sessions, he was negotiating with U.S. Secretary of State Frederick Frelinghuysen (who attended only some of the IMC sessions).[24] In the midst of a divisive presidential election campaign, the United States was a ripe source of fascination for Galvan and other diplomats, including the British envoy, Sackville-West. They had much to report back home about the election's possible ramifications for their own countries' relations with the United States.[25]

Figure 3.2 | IMC delegates, 16 October 1884, in photo taken on front steps of Executive Office Building (where conference was held), after reception at White House, across the street, hosted by U.S. president Chester Arthur.

Front row (*left to right*): Alvensleben (Germany), Galvan (San Domingo), Lefaivre (France), large gap before Shaeffer (Austria) at far right.
Second row (*left to right*): Hinkeldeyn (Germany), Valera (Spain), Soteldo (Venezuela), Strachey (Britain [India]), Janssen (France) (posing dramatically with his back to Adams and Rodgers), Rodgers (USA), Cruls (Brazil), Weckherlin (Netherlands); Third row (*left to right*): Frey (Switzerland), Kikuchi (Japan), Rutherfurd (USA), Franklin (Colombia), Evans (Britain), Adams (Britain), W.F. Peddrick ([USA] reporting secretary), Rustem Effendi (Turkey) (standing slightly further forward than Peddrick), Foresta (Italy). Fourth row (left to right): Pastorin (Spain), Abbe (USA) (partially hidden by Rutherfurd's head), a small gap before Kologrigoff (Russia) and de Struve (Russia) (both with upside-down, V-shaped beards), Stebnitzki (Russia), Fernandez (Mexico) (slightly in front and right of Stebnitzki), Sampson (USA), Anguiano (Mexico), A.A. Adee (U.S. ass't secretary of state) (standing facing left towards Anguiano); Back row (*left to right*): Gormas (Chile), Tupper (Chile), small gap before Lewenhaupt (Sweden), Batres (Salvador), Echeverria (Costa Rica) (slightly behind and right of Batres), Miles Rock (Guatemala), Alexander (Hawaii) (slightly behind and right of Rock), Aholo (Hawaii), Stewart (Paraguay), F.R. Marceau (French-language stenographer).
Absent: Allen (USA), Arbol (Spain), Coppinger (Liberia), Fleming (Britain [Canada]), Hirsch (Switzerland).]

Many of the diplomats at the IMC were assigned to Washington long term and had established relationships that the new arrivals lacked. They made connections with local elites and were the subject of both admiration and gossip. *Hon. Uncle Sam*, a tongue-in-cheek exposé of Washington high society, described the social rankings of various diplomats. Sackville-West (later Baron Sackville) ranked at the top. "I am inclined to think Uncle Sam dearly loves a lord – when he comes from England,"[26] the book chided. The French minister was less well respected. "Is it because he has no title? Is it because he is a republican?" the author wondered.[27] Meanwhile, "Baron de Struve, the Minister from Russia, is popular in society. The Americans rather like the Russians." The book gave high marks to de Struve, who represented Russia at the IMC. Although he spoke several Asian and Slavic languages, he had not yet mastered English.[28] American astronomer Simon Newcomb, in a letter to Russian astronomer Otto Struve, the ambassador's brother, wrote: "You would be delighted to know what your brother has done for the Russian legation here. I think he is the most popular of all the foreign ministers [*sic*] in Washington and is sought after on all occasions."[29]

Hon. Uncle Sam also praised the Italian delegate to the IMC, Albert de Foresta. To dance with him, it said, "is an honor that any girl will remember all her life."[30] Clearly, this was a social network as much as a professional one. American socialite Marian Hooper Adams made friends with the wife of the Swedish minister and IMC delegate, Count Lewenhaupt.[31] Their activities were followed by newspapers and by more esoteric publications alike. The *Phrenological Journal and Science of Health*, for example, made dubious declarations on the disposition of the various ministers based on their skull shape. Sackville-West had "fine manners" and "a practical mind," while German IMC delegate Baron H. von Alvensleben was "a man of quick mental impressions, and rather earnest and intense in feeling."[32]

Not everyone was so closely connected to these social networks. Fleming, for example, tended to his business endeavours more than diplomatic or scientific networking. Between IMC sessions, he left Washington several times. In the week between the third and fourth sessions, he travelled to New York, where he met with Donald Smith and George Stephen, the leading promoters of the Canadian Pacific Railway, as well as Conservative Prime Minister Sir John A. Macdonald, who was on his way to England.[33] The four of them discussed the building of the St Lawrence Bridge in Montreal. Jeannie, Fleming's wife, went back to Montreal with Donald Smith, riding in his private car named "Saskatchewan." Fleming returned to Washington alone.

Fleming left Washington again after the fifth session for another week, this time to Montreal to deal with some Hudson's Bay Company matters (he was a company director), as well as engaging in more discussion about the St Lawrence Bridge.[34] Meanwhile, the rest of the IMC delegates were courted with a presidential reception. On 16 October, they gathered at the White House and met Republican President Chester A. Arthur. After a few speeches and a round of hand-shaking, they were given a tour of the White House, before being paraded outside for a group photograph.[35] Afterwards, British delegates Professor J.C. Adams and Sir Frederick Evans dined at the Cosmos Club with Commodore S.R. Franklin, the American naval astronomer who was representing Colombia at the IMC. Unfortunately, there is no telling what they said. The next day, 17 October, the delegates again rubbed shoulders with the political elite, this time with a cruise down the Potomac to Mount Vernon with Secretary of State Frelinghuysen.[36] Fleming, busy in Montreal, was not there. Of course, he had done the trip to Mount Vernon before, but not with this company.

These receptions, dinners, and outings were specifically designed for IMC delegates to socialize in both formal and informal settings. They created a rather strange intermingling of networks, as Washington high society mixed with international astronomical experts. Scientists, diplomats, and businessmen alike were given the chance to interact beyond the confines of the IMC's elegant meeting hall. But each prioritized his own network. Fleming met with his railway contacts; Adams and Kikuchi attended academic gatherings; Galvan workshopped his trade agreement.

Clearly the people at the table at the IMC were arriving with very different preconceptions. Diplomats looked at the event in terms of national interest; astronomers and naval officers regarded it as a sequel to the Venice and Rome conferences, seeking a prime meridian for navigation; while businessmen and engineers like Fleming and Allen had civil timekeeping in mind, based on their careers in commerce and railway management. The proceedings in the grand hall of the State, War and Navy Building were in large part a result of what went on outside of it. The fault lines between delegates in session at the IMC closely matched pre-existing professional differences.

PRECONCEPTIONS

What expectations did delegates have of the IMC at its start (see official seating chart in Figure 3.3)? As we have seen, Fleming arrived understanding that the conclave was tasked with establishing standard

Country	Name	Seat	Desk	Seat	Name	Country
			President Rear Admiral C.R.P. Rodgers	1		
États-Unis	Professor Cleveland Abbe	41		2	Count Carl Lewenhaupt	Suède
	Commander W.T. Sampson	40		3	Baron Ignatz Von Schaffer	Autriche-Honorie
	Mr. W.F. Allen	39		4	Dr. L.Cruis	Brésil
	Mr. Lewis M. Rutherford	38		5	Mr. F. V. Gormas	Chilé
Venezuela	Senor Dr. A.M. Soteldo	37		6	Mr. A. B. Tupper	Chilé
Turquie	Mr. Rustem Effendi	36		7	Commodore S. R. Franklin	Colombie
Suisse	Proffesssor Hirsch	35		8	Mr. J. F. Echeverria	Costa Rica
	Col. Emile Frey	34		9	Mr. A. Lefaivre	France
Espagne	Mr. Juan Pastorin	33	Desk	10	Mr. Janssen	France
	Mr. Emilio Ruiz Del Arbol	32		11	Baron H. Von Alvensleben	Allemagne
	Mr. Juan Valera	31		12	Mr. Hinckeldeyn	Allemagne
Salvador	Mr. Antonio Batres	30		13	Captain Sir F. J. O. Evans	Grande-Bretagne
Saint Domingue	Mr. M. de J. Galvan	29		14	Professor J.C. Adams	Grande-Bretagne
Russie	J. de Kologrivoff	28		15	Lieutenant General Strachey	Grande-Bretagne
	Major-General Stebnitzki	27		16	Mr. Sandford Fleming	Grande-Bretagne
	Mr. Chas. de Struve	26		17	Mr. Miles Rock	Guatemala
Paraguay	Captain John Stewart	25		18	Honourable W. D. Alexander	Hawaii
Hollande	Mr. G. de Weckherlin	24		19	Honourable Luther Aholo	Hawaii
Japon	Professor Kikuchi	23		20	Count Albert de Foresta	Italie

22	21
Mr. Angel Anguiano	Mr. Leandro Fernandez
Mexique	

Figure | 3.3. IMC seating chart.

time worldwide, based on a new common prime meridian. It is easy to understand why he felt this way: this was *his* conference. The IMC would never have taken place if not for his lobbying. What is more, a common prime meridian was unnecessary for navigation. Any line of longitude would do. (More than a dozen were currently in use. A nuisance, but not necessarily a problem, except in extreme circumstances, as William Parker Snow illustrated.) A shared system of timekeeping, in contrast, required a prime meridian. It made perfect sense that the IMC should therefore focus on timekeeping and not on longitude.

Not everyone shared these priorities. The rest of the British delegation arrived with entirely different expectations. The official letter of instruction received by Professor Adams, Sir Frederick Evans, and General Sir Richard Strachey had two main points. First, Britain was not to be bound to any decision, especially if it should declare anything other than Greenwich a prime meridian. Second, the delegates were to avoid in every way possible any discussion of the metric system. The three men were given no guidance as to how they should deal with time reform. The subject was simply unimportant to the Science and Art Department, which had appointed them.[37]

Adams outlined his own preconceptions in a letter to American astronomer Simon Newcomb in July 1884: "I hope that the business will be strictly confined to the question of the Prime Meridian, and the time from which the astronomical day shall be reckoned, and not allowed to be mixed up in any way with other questions."[38] Although Adams did not specify what the other questions might be, context suggests that he meant two things: standard time and, probably more important for him, the adoption of the metric system. Adams did have an interest in timekeeping, but it concerned only the astronomical day as used by astronomers, not the civil day used by everyone else. He had no wish to alter the way ordinary people measured time. Adams was coming to the IMC with a very narrow view of its aims and scope.

Some of Fleming's allies recognized this difference in opinion and tried to warn him. Barnard wrote to him in late September. Having just decided to resign his seat, he wanted to ensure that everything would go smoothly in his absence. His primary concern was reining in Fleming's ambitions. He asked the engineer to consider carefully the conference's scope: "I am of the opinion," he wrote, "that it would be best not to go beyond the single object for which the conference is ostensibly called viz. the agreement upon a common Prime Meridian, and that dependant questions which may naturally grow out of this should be put aside."[39] Although he thought standard time a fine goal, a discussion of it might not go their way. Any disagreement "might endanger [sic] a failure in regard to the main object."[40] He also warned his friend that raising the idea of a twenty-four-hour day, or other calendar reforms, might have similarly disastrous consequences.

Fleming did not take the hint. Barnard warned him again on 2 October: "I am afraid … that you are disposed to occupy the Conference with the uses to which the Prime Meridian may be put, in advance of the determination of the more important, and only really essential question, what shall the Prime Meridian be."[41] If Fleming were to bring up the meridian's uses, such as standard time, "differences of opinion are liable to arise which may not easily be adjusted, and which, if brought prematurely into activity, may prejudice the main question."[42] Barnard begged his colleague to leave aside time reform until after a meridian had been chosen. But this advice, as we see below, went unheeded.

While most histories of the IMC have focused on the French-British conflict over the prime meridian's location, the real struggle was over its uses. Was it for determining longitude alone, and perhaps for astronomical timekeeping, as Adams desired, or should it enable a global reform in civil timekeeping, as Fleming wanted? Lines were drawn not between nations, but between competing professional networks.

THE FIRST DAYS

This state of affairs was not obvious on day 1 (Wednesday, 1 October 1884). Indeed, very little happened at all, besides the formalities of introductions, and the election of a conference president (Admiral Rodgers). There was some discussion about the languages to be used in the conference proceedings (they agreed on French and English), and whether the public might be allowed to attend, but this last question was put off to the second session.

It was in that second session (Thursday, 2 October) that fault lines began to emerge. Career diplomats like French delegate A. Lefaivre, for example, seemed flustered at the naïveté of some of the scientists about how diplomatic meetings were run. British and American delegates had proposed a motion to invite other scientific experts to participate in the deliberations. But these experts were not authorized to speak on behalf of any government, Lefaivre protested. "It was not in accordance with the object of this Conference that private individuals, not authorized by their respective Governments, should be permitted to influence the decision of this body."[43] Ultimately, the experts were invited, but not as full participants. They could speak only when called upon. A similar proposal, which would have allowed the public to participate, was also rejected by Lefaivre, and it was further decided not to answer any of the correspondence or proposals received from the public by mail. This was to be a closed and guarded process, not an open forum.

The IMC formed a committee to read and summarize any correspondence received and to recommend whether to pursue any of the letters' propositions. British delegate J.C. Adams, the chair of this committee, warned that "the Conference should be very cautious in admitting the devices and schemes of people who have no connection with this body; that there are, no doubt, many inventors and many people who have plans and schemes which they wish to press upon the Conference."[44] Certainly he was right about some of the letters. A few consisted of time-reform schemes based on new inventions or clock dials, for which the writer had the patent and was hoping to cash in. Several suggested Greenwich as prime meridian, or 180 degrees from it. Bethlehem and Jerusalem were both proposed on religious grounds, with writers noting that the western calendar was based on the birth of Christ. The writer of the Bethlehem proposal brought up William Parker Snow's choice of the St Paul's Rocks, praising Snow's effort to pick a neutral location to avoid national jealousy. But ultimately he preferred the sacred symbolism of Bethlehem to the lonely rocks in the Atlantic, which Snow had proposed. Other writers also tried to ease the expected French-British tension. A French author supported a neutral prime meridian, while another suggested Greenwich, but calling it Le Havre (a French town on the same meridian as Greenwich).

The desires of Scottish astronomer royal Charles Piazzi Smyth made their way into the correspondence via the IPAWM, whose members sent multiple letters and pamphlets recommending the Great Pyramid of Giza as prime meridian and expressing their fear that Barnard and the French delegates would force the metric system on the world. One of the more colourful proposals came from a patriotic American who nominated the "pristine shaft" of the soon-to-be-completed Washington Monument as the prime meridian, given the symbolism of the obelisk as the highest achievement of humankind.[45]

While some proposals were rightly dismissed outright, especially those with patents attached to them, others were reasonable suggestions. Unfortunately, the committee did not report on the correspondence until the sixth session, on Monday, 20 October, by which time a prime meridian had been chosen, rendering the majority of the letters useless. The insular nature of diplomacy silenced outside opinion in favour of those sitting at the table. As a result, Snow and Smyth's opinions made it to the table at Washington only belatedly.

The second session (Thursday, 2 October) of the IMC was relatively short. After the decision about public correspondence was made, the delegates moved on to other matters. The anticipated British-French rivalry began to rear its head. The French diplomat Lefaivre asserted that the IMC was not binding on his country, but must simply recommend. He also took pains to disavow the resolutions of the Rome Conference (which had decided on a Greenwich meridian), stating that this current gathering was different: it was to look at political, not just technical, variables.[46] He was setting the stage for a rigid opposition to Greenwich.

It was also during this second session that it was decided to vote by nation, rather than by individual, removing Fleming's ability to cast votes independently of the rest of the British delegation. The conference adjourned for its second day, having agreed on only one resolution: that a single prime meridian was more desirable than the "multiplicity" of meridians that currently existed.

DAYS 3 AND 4: DRAWING LINES IN THE SAND

There were a few days of recess before the conference met for the third time, providing delegates with an opportunity to plan and to meet with possible allies. On Friday, 3 October, Fleming dined with one of the Spanish delegates, Juan Pastorin.[47] The two had corresponded on time reform in the past, and Pastorin was a supporter of Fleming's ideas for reforming civil time. Now they were together with the opportunity to implement their schemes. But the two men both faced a similar challenge. Their opinions on time reform differed from the rest of their own delegations. Fleming and his fellow British delegates, especially Adams, agreed on almost no aspect of the IMC, and Pastorin found that his Spanish colleagues were equally disinclined to support standard time. Indeed, Fleming, Pastorin, and railwayman William Allen were the only delegates at the conference who showed any interest in the subject (Cleveland Abbe, strangely, said next to nothing about timekeeping, perhaps taking to heart Barnard's advice about focusing on the main object of choosing a prime meridian, counsel Fleming had ignored). The rest of the delegates were concerned with a prime meridian for navigation, and perhaps a universal day for astronomical purposes, but little else.

After meeting with Pastorin, Fleming spent the rest of his downtime drafting a notice to his fellow delegates, which he circulated on Saturday, 4 October. His note urged them to consider time reform alongside the longitude problem, rather than separating them. He also sent around a pamphlet on uses of standard time, hoping to spur some discussion, and at the very least show his fellow delegates that the prime meridian was related as much to timekeeping as to longitude.[48]

As the third session opened on Monday, 6 October, it became clear that Fleming's pleas had fallen on deaf ears. Instead, the session opened the floodgates of the Anglo-French rivalry. Lengthy speeches from both sides argued back and forth about where to place the prime meridian, with

varying grades of eloquence. France, resting its case on the principle of
scientific neutrality, argued that the prime meridian should not lie within
any national boundary, nor even intersect "any great continent – neither
Europe nor America."[49] The Bering Strait or an Atlantic isle seemed the
most advantageous. Britain and the United States, meanwhile, invoked
two points: convenience and scientific precision. Most of the world's
shipping already used Greenwich, making it the most convenient choice
for navigators. Furthermore, longitudinal determination required precision that was possible only with a top-notch observatory, such as Paris,
Greenwich, Berlin, or Washington. An imaginary neutral line in the middle
of an ocean would be useless for precise calculations, they insisted.[50] The
argument between the two sides became so contentious that no vote was
taken that day, and the conference adjourned with unfinished business.

   This debate between France and Britain has been thoroughly analyzed
elsewhere, and I will not repeat it here.[51] But one small interjection made
early on day 3, easily missed amid the national rivalries, illuminates the
place of timekeeping in the debate. Back on day 2, American astronomer Lewis Rutherfurd had tabled a proposal to adopt Greenwich
as the prime meridian, a move that began the French-British fracas.
However, before the motion was discussed on day 3, he amended it to
read: "That the Conference proposes to the Governments here represented the adoption of the meridian passing through the centre of the
transit instrument at the Observatory of Greenwich as the standard
meridian for longitude."[52] The only real change was the addition of the
last two words: "for longitude." The initial proposal had not referred
to the uses of the new prime meridian. Now, it was made explicit. The
prime meridian was to be a tool for establishing longitude, whether
for navigation, surveying, or mapmaking. This definition purposefully
excluded timekeeping.

   The conference proceedings suggest that the amendment to include
the words "for longitude" was passed unanimously, but it is hard to
imagine Fleming acquiescing so easily to a motion that threatened his
entire purpose in campaigning for the previous five years to bring about
the IMC, and completely undermined the memo he had just circulated to
all the delegates two days earlier. Unfortunately for him, because of the
change in voting procedure, his vote was now subject to the whims of
the other British delegates, none of whom was interested in time reform.

   This was the first hint of disaster for Fleming's plan, and it makes no
sense if we consider the rivalries at the conference only along national
lines. It makes much more sense if we look at the delegates by occupation. The railway engineers and Pastorin were the only delegates at the

table committed to time reform. The astronomers and naval officers in attendance (Pastorin excepted) were concerned only with longitude. This professional divide would only become clearer in the coming weeks.

A six-day recess (Tuesday, 7–Sunday, 12 October) followed the contentious third session. This was the week most heavily reported by the press, which was fascinated by the Anglo-French antagonism. "A Strong Probability That No Agreement Will Be Reached," proclaimed a *New York Times* headline on 8 October.[53] "It is believed that the conference will fail," the article continued. "The opposition of the French delegates to the adoption of an English meridian is still firm, although that opposition seems to be the result of patriotic and sentimental rather than partisan considerations."[54] The British and American papers were unsurprisingly biased against the French position. "Should the conference fail in the object for which it has met, the responsibility will rest entirely on France," wrote the *Daily News*.[55] Of course, the British delegates were equally stubborn, but the Anglo-Saxon papers did not say so.

News of the impasse spread far and wide. An American acquaintance wrote to Sir Richard Strachey on 11 October, inviting him to New Orleans when the conference finished. He hoped "our testy French cousins may become more reasonable in their demands, agree upon Greenwich as the only proper common meridian, and thus enable you to see something of our country before you return to England."[56]

Of all the delegates, the French astronomer Jules Janssen faced the most intense scrutiny from the newspapers that week. Their pestering, along with a sweltering heat wave, left him in a state of exhaustion. That October began with unseasonably warm temperatures. On Saturday, 4 October, Fleming recorded in his diary that it was over 90 degrees Fahrenheit.[57] The gilded diplomatic hall in the State Department was stuffy and uncomfortable, providing little relief from the temperature outside. In late October, the weather then swung violently the other way: by the seventh session on Wednesday the twenty-second, Fleming recorded, it had now become "very cold."[58]

During that gruelling first week, Janssen had struggled to stay positive. Each session, he faced an onslaught of verbal abuse, as the English-speaking delegates, one after another, tried to tear down his proposal for a neutral meridian (Janssen, who spoke only French, received notes from Lefaivre explaining their counterarguments). The heat just made it worse, as Janssen explained to his wife: "I fought for four hours tirelessly. When I left this boiler-room, my shirt was wringing wet: it took two days to dry … And because here everything is done with breathtaking rapidity, I had to spend two days without getting undressed …

Added to that, the whole of the American press is hot on our heels every day. You can imagine whether it is a bed of roses."[59]

Others, not so hard-pressed, found ways to escape the humdrum. Sundays were days of rest but also allowed opportunities for a form of religious tourism. Fleming noted in his diary on Sunday, 5 October that he attended a Black Presbyterian church, followed by a second service elsewhere in the afternoon.[60] Similarly, a week later, he invited Juan Pastorin to join him at a Presbyterian service, the Catholic Spaniard's first time in a Protestant church.[61]

But Fleming did not waste the opportunity to prepare either. He needed to try again to refocus the conference away from squabbles arising from national pride to focus instead on timekeeping. After the week-long recess, delegates reconvened on Monday, 13 October. The fourth session began with some housekeeping, but after that was finished, Fleming was the first to speak. In an attempt to rescue what seemed to be a failing cause, he had prepared a lengthy speech. He was hoping to establish a system of global civil timekeeping, similar to what he had already introduced to the United States and Canada. This timekeeping system *required* a single prime meridian. Navigation did not. After the conference, all the other delegates' countries could, if they wished, leave the conference and resume using their national meridians for navigation, with little consequence. But for Fleming's timekeeping plan, that would be disaster. So he needed to bridge the gap between the French and British and ensure that a prime meridian was chosen.

He proposed a compromise. The prime meridian should be placed in the middle of the Bering Strait – neutral, as France proposed. But it would also be exactly 180 degrees from Greenwich, allowing the precise calculations that astronomers required to be made at Greenwich Observatory, and then easily translated to the prime meridian by a simple calculation. Of course, the association with Greenwich hurt Fleming's claim of "neutrality," but it was nevertheless his intention to reconcile the two opposing sides.

His compromise fell on deaf ears. The Brazilian delegate, Luis Cruls, immediately doubled down on the principle of neutrality proposed by France. Cruls was Belgian; he had moved to Brazil in 1874 to work at the Imperial Observatory in Rio de Janeiro.[62] He was trained by a French astronomer, Emmanuel Liais, which might have biased him towards the French proposals. The ongoing conflict of Emperor Dom Pedro II with Britain over the legality of the slave trade may also have influenced his decision.[63] Furthermore, Brazil was not one of the 72 per cent of countries that already used Greenwich for navigation, instead using Rio de

Janeiro. It had no reason therefore to support the British position.[64] Unfortunately for Brazil, it was in the minority. As Jules Janssen later complained, the United States had invited a horde of smaller countries as allies to overwhelm any opposition.[65]

Fleming's compromise satisfied no one. Brazil and France continued to press for a neutral meridian, and nearly everyone else advocated for Greenwich. When a vote was taken on the principle of neutrality, it was swiftly defeated. Only France, Brazil, and San Domingo had voted for it. With this barrier out of the way, the American delegate Lewis Rutherfurd immediately reintroduced his proposal to adopt Greenwich as the prime meridian for longitude.

Seeing his compromise slipping away, and desperate for a way to reintroduce timekeeping back into the debate, Fleming proposed his own amendment to the resolution: "That a meridian proper, to be employed as a common zero in the reckoning of longitude *and the regulation of time* throughout the world, should be a great circle passing through the poles and the centre of the transit instrument at the Observatory of Greenwich."[66] The amendment did two things. First, it put timekeeping front and centre as one of the primary purposes of the prime meridian, alongside the determination of longitude. This was an attempt to undo the damage caused by Rutherfurd, whose earlier amendment had mentioned only longitude, ignoring timekeeping. Second, Fleming carefully phrased it to keep his compromise on the table. If the meridian was a great circle, then it could still be interpreted to mean that the neutral Bering Straight, 180 degrees from Greenwich, and not Greenwich itself, was the meridian. Fleming still needed to hold out the olive branch to France if his global timekeeping system was to succeed.

What happened next is one of the key moments in which a focus on national rivalries rather than occupational ones is particularly unhelpful in explaining delegates' actions, because Fleming, a representative of the British Empire, was immediately undercut by his fellow representatives of the British Empire. The Cambridge astronomer J.C. Adams stood up to quash Fleming's amendment. "I desire merely to state ... that the remaining delegates of Great Britain are by no means of the opinion expressed in that amendment, and that it is their intention, if it should come to a vote, to vote against it."[67] The SAD had earlier feared that Fleming's opinions might not align with Britain's interests. Those fears proved well founded. Fleming's amendment was rejected by Adams, and further condemnations followed. The German delegation suggested that the amendment mixed up two questions, and that time and longitude should be considered separately.[68] When the amendment was put to a vote, it lost, decisively.

The conference fell back into national camps. Spain revived the compromise proposed at Rome whereby France would adopt a Greenwich meridian and Britain would go metric. This was quickly stymied by both sides. Lefaivre, still clinging to scientific neutrality, argued (somewhat prophetically) that if Greenwich were chosen out of convenience, because it currently was the meridian most commonly used by commercial shipping, it would soon seem an anachronism: "Nothing is so transitory and fugitive as power and riches. All the great empires of the world, all financial, industrial, and commercial prosperities of the world, have given us a proof of it, each in turn."[69] Britain's dominance of the seas, in other words, would not last forever.

Lord Kelvin, one of the non-voting experts invited to the conference, was called on to respond. He reiterated, correctly, that no one meridian was more "scientific" than another. Any line of longitude would do. But Greenwich, he argued, ignoring Lefaivre's warning, was the most convenient.

When Rutherfurd's resolution was finally voted on, only San Domingo opposed it; France and Brazil abstained. One London publication suggested that Galvan from San Domingo did not understand the question at hand.[70] Such claims are unfounded. Galvan, like the French delegates, wanted a neutral meridian, and a neutral meridian in mid-Atlantic would be closer to San Domingo local time than Greenwich.[71] But the majority had spoken: Greenwich was now the world's prime meridian.

Only now, with the question of a prime meridian for longitude settled, were participants willing to discuss timekeeping. Fleming found a surprise ally. The Russian delegate, Charles de Struve, happened to be the half-brother of Otto Struve, a Russian astronomer who agreed with some of Fleming's time-reform ideas. In particular, the Russian delegate supported the introduction of a twenty-four-hour clock to replace twelve-hour clocks. He also recommended a universal day, based on the prime meridian, but did not fully endorse Fleming's radical reform of civil timekeeping. De Struve proposed not to change local time, but simply to introduce a separate universal time "for international telegraphic correspondence, and for through international lines by railroads and steamers."[72] In essence, he was advocating two tiers of time: universal time for specialized tasks, and local time for civil and everyday use. This was not what Fleming wanted.

The rest of session four devolved into a discussion about which way to count longitude: east and west up to 180 degrees, or in one direction 360 degrees. The conference adjourned without an answer. It had been a day of mixed results for Fleming. He had his prime meridian now, but France was still noncommittal, and civil-time reform was still not up for discussion.

DAY 5: PUTTING TIME REFORM ON THE TABLE

At the IMC's fifth session, held the next day, Tuesday, 14 October, Fleming was again the first to speak, aside from a few formalities. His goal was to turn the delegates' attention away from the relatively minor question of which direction, east or west, to count degrees of longitude, and direct them instead towards the larger task of standardizing global timekeeping. "To my mind," he began, "longitude and time are so related that they are practically inseparable, and when I consider longitude, my thoughts naturally revert to time, by which it is measured. I trust, therefore, I may be permitted to extend my remarks somewhat beyond the immediate scope of the resolution."[73] In the space of a few generations, Fleming observed, "the application of science to the means of locomotion and to the instantaneous transmission of thought and speech have gradually contracted space and annihilated distance. The whole world is drawn into immediate neighborhood and near relationship."[74] The lack of a common time in this contracting world was bound to become more and more of a nuisance, if not dealt with now. What was needed was a single universal day, based on the time at the prime meridian. Fleming appealed to the scientific sensibilities of the room, suggesting that a myriad of local times was "inconsistent with reason." But he also spoke to the general utility of a shared timekeeping system: it should be tailored towards practical applications. Convenience, in other words, was as important as precision. Ordinary people were used to rising at seven, eating lunch at noon, and so on. A single universal day would mean that some people would rise at midnight, have lunch at breakfast, and so on, depending on where they lived around the globe. This change would be something of a shock to many. So a compromise was needed between local and universal time. Standard time, as had been implemented in the United States, was the solution, the tool that linked local time to universal time in a rational way.[75] It created twenty-four local times, all one hour different, and in line with universal time, to replace the thousands of local times currently in use.

To simplify the connection between time and longitude, Fleming suggested that longitude be counted in one direction around the globe. If one imagined the turning earth like a clock face with twenty-four hours, longitude and time reckoning would be united in perfect harmony.

When he finished speaking, Fleming's impassioned appeal was immediately undercut, once again, by his fellow delegates from Great Britain. Professor J.C. Adams rose to suggest instead that longitude be counted in two directions, east and west. He similarly dismissed standard time zones, suggesting instead to continue using local time, which could be determined by using a simple formula: "Local time at any place is equal to universal time plus the longitude of the place ... Now, I think it is perfectly impossible for Mr. Fleming to make a more simple formula than that."[76] Adams's simple formula, were it put into practice, would entrench local civil time for civil use and exclude the use of time zones.

Sir Frederick Evans of Britain also dismissed Fleming's plan. He asked the delegates to disassociate the time question from longitude, looking at it from the point of view of a navigator. Longitude at sea, he argued, was always counted in two directions, not one. Evans continued: "My colleague, Mr. Fleming, made the remark that he could not disassociate longitude from time. If he had mixed with seamen, he would have found out that there is very frequently a well defined difference between the two in their minds. Longitude with seamen means, independently of time, space, distance. It indicates so many miles run in an east or west direction. Consequently, I am not able to look upon longitude and time as being identical."[77] There was a fundamental disagreement between professions about the very nature of longitude and its relationship to time. Engineers like Fleming, and astronomers like Adams, came to the IMC with very different goals. Evidently the opinions of naval officers were split. Pastorin sided with Fleming, but all the other navigators in the room, including Evans, opposed him. There is little room for other interpretations when Evans told the IMC that to him universal time was "a matter of indifference."[78] While for Fleming the whole purpose of the IMC was timekeeping, for astronomers and most navigators, timekeeping was immaterial, and longitude was the only real question up for debate.

The fourth and final British delegate (for India), Sir Richard Strachey, spoke next and attempted to find some common ground between his own delegation's two warring factions. He agreed with Adams and Evans on the counting of longitude in two directions, but he also did not dismiss Fleming's plea for a universal day. Instead, he proposed that the international date line (not yet called that, where the new day begins) be 180 degrees from Greenwich, so that the date changed at Greenwich midnight, rather than noon.

No one took up Strachey's proposal right away. Instead, the conference proceeded to vote on the direction of counting longitude. Counting both directions from Greenwich, east and west, won out. France, Brazil, and San Domingo, among others, abstained, still refusing to recognize the authority of a Greenwich meridian in the first place.

After the resolution passed, universal time was at last put on the table for formal discussion. The proposed resolution read: "Resolved, That the Conference proposes the adoption of a universal day for all purposes for which it may be found convenient, and which shall not interfere with the use of local time where desirable."[79] One phrase stands out: "Shall not interfere with the use of local time." This phrase earmarks this resolution as one that aligned more with Adams than with Fleming. The universal day as envisioned here was to be a tool for specialized purposes, not for general adoption by the public. Local time would still govern everyday life.

The resolution was immediately controversial, from both sides. The Italian delegate proposed as an alternative the resolution adopted at Rome: "The Conference recognizes, for certain scientific needs and for the internal service of great administrations of ways of communications, such as those of railroads, lines of steamships, telegraphic and postal lines, the utility of adopting a universal time, in connection with local or national times, which will necessarily continue to be employed in civil life."[80] This was in its essence the same resolution, except that it even more explicitly separated universal time for "scientific needs" and local time for "civil life." Before either resolution could be discussed, William Allen proposed a third alternative, which swung the emphasis the other way: "Civil or local time is to be understood as the mean time of the approximately central meridian of a section of the Earth's surface, in which a single standard of time may be conveniently used."[81] Allen's amendment proposed standard time for civil use, such as the system American railways adopted the previous year. The floodgates were finally open to discuss timekeeping, and it was divisive. At the centre of the controversy: should the universal day apply to everyone via standard time, or be for scientific purposes only, with local time still informing daily life?

Allen framed his resolution as a protest against the creation of a hierarchy of time systems. Businessmen, scientists, railways, and common people, he said, must all share the same time system. Standard time zones, he argued, were the best way to satisfy the needs of everyone. He went on to illustrate the success of standard time in North America, as testament to its utility. Allen's resolution was the most cogent argument for standard time yet voiced at the conference, and the closest it came to becoming a reality. But its moment in the spotlight was destined to be brief. In a strange about-face, Allen almost immediately withdrew his resolution, possibly due to Rutherfurd's objection that defining local time was beyond the conference's purview.[82]

The Italian resolution was defeated soon afterwards. Delegates felt it too specific, preferring the vague wording of the original resolution, which left the uses of universal time open to whatever interpretation might be desired. As a compromise, the wording of the original resolution was tweaked: the universal day "shall not interfere with the use of local *or other standard time* where desirable."[83] The added words allowed a nation to choose its own method of timekeeping – national, local, or standard. Instead of establishing standard time across the globe, as Allen's resolution would have done, the IMC adopted a resolution that allowed each country to choose its own method of civil timekeeping. It was approved unanimously (Germany and San Domingo abstained). The universal day now existed, but was restricted in its use, subordinate to whatever local timekeeping system a nation wished to adopt. The conference had decided that standard time was an option, not an obligation.

Nonetheless, a universal day had been established. This was no small thing. If we keep in mind Ogle and Nanni's insight that global time reform was a Europeanizing project that reinforced colonial realities, the unanimous adoption of a "universal" day should give us pause. While few colonial voices were present to object (Fleming and Strachey alone represented any "colony"), there were some non-western nations represented. The response of the Ottoman Empire's delegate, Rustem Effendi, offers a glimpse at the shallow "universality" of the universal day.

Rustem was the son of a Polish refugee who escaped Poland's failed 1848 revolution and moved to the Ottoman Empire in 1854. Born on the Ottoman island of Midilli, he was originally Alfred Bilinski, but he changed his name upon conversion to Islam.[84] Rustem joined the Ottoman diplomatic service and in 1884 was the empire's envoy in Washington, speaking Turkish, French, English, and Italian fluently.[85]

Ottoman reformers in the 1880s were painfully aware of the threat the west posed to their empire, particularly after Britain's occupation of Egypt.[86] And, just as Britons were concerned about "national decline" in the face of other great powers, Ottoman reformers similarly feared decline. Throughout the 1880s, Ottoman intellectuals aimed to revitalize their people and stave off external threats by instilling an ethic of self-improvement. This ethic included notions of time management and disparaged its opposite, time wasting.[87] In practical terms, meanwhile, the Ottoman Empire balanced a multiplicity of times, including differing religious and legal forms of timekeeping. Rustem was thus representing a vibrant and complex polity, whose interests were somewhat different from those of his colleagues.[88]

Rustem voted in favour of a universal day. However, he had made clear that his government was not bound to any decision of the conference.

This was true of all delegates, but Rustem was emphatic: "My vote does not bind my Government. I am, indeed, obliged to vote against any proposition which would tend to bind it in any way, for I desire to leave it free to act in the matter."[89] After the universal day passed, the conference was preparing to move on to its details: when would the day begin, for example? But Rustem was not ready to accept the universal day in principle just yet.

Speaking candidly, Rustem undermined some of the assumptions that underpinned the notion of a universal day. "The question of a universal hour is not of equal interest and importance to all," he began.[90] Smaller countries, he explained, could make do with a national hour. The complex systems required in the sprawling United States, Canada, Russia, or the British Empire were of little use to France, Japan, or Italy, for example. As for the Ottoman Empire, it too had little need for another method of timekeeping. Indeed, the Ottomans required "more latitude" (as it were) than other nations with respect to the universal day. Rustem explained:

> In our country we have two modes of reckoning time: one from noon to noon, or from midnight to midnight, as everywhere else, (*heure à la franque*), the other (*heure à la turque*) from sundown to sundown. In this latter case the hours count from the moment when the disk of the Sun is bisected by the horizon, and we count twice from oh. to 12h., instead of counting without any interruption from oh. to 24h. We are well aware of the inconveniences this system of counting produces, because oh. necessarily varies from day to day, for the interval of time between one sunset and the one following is not exactly 24 hours. According to the season the Sun will set earlier or later, and our watches and clocks at Constantinople will be at most about three minutes fast or slow from day to day, according to the season.
>
> Reasons of a national and religious character prevent us, however, from abandoning this mode of counting our time. The majority of our population is agricultural, working in the fields, and prefer to count to sunset; besides, the hours for the Moslem prayers are counted from sundown to sundown. Therefore it is impossible for us to abandon our old system of time, although in our navy we generally use the customary reckoning or *"heure à la franque."*[91]

Rustem wanted to ensure that the universal day would be used for only international affairs and that it would not interfere with his country's domestic timekeeping practices. His hesitation was based in religion, and he apologized for its being not scientific in nature, but of "a different and inferior order."[92] His apology should have been unnecessary. After all, he was far from the only one to make claims about measuring time vis-à-vis religion. The Spanish delegates attempted to establish Rome as the anti-meridian for global timekeeping, based on the Gregorian calendar. Piazzi Smyth (not a delegate) had nominated the Great Pyramid of Giza as the prime meridian for religious reasons, like others' proposals for Jerusalem and Bethlehem. Christian sensibilities were just as prominent in the time-reform debate as Islamic ones.

Rustem's objection met with little discussion. It was pointed out to him that the resolution already conceded that universal time would not in any ways interfere with local time.[93] And so the universal day entered into existence. Now it was time to define it.

The first order of business was to determine when the universal day should start (essentially, where should the international date line be?). The Spanish diplomat in the room, Juan Valera, motioned to leave the discussion for the next day. After all, he said, he had already fulfilled his mandate: to select a prime meridian. He felt this discussion about timekeeping was beyond what his government had authorized him to do, so he needed time to deliberate. The session adjourned.

DAY 6

The conference delegates had a hard time agreeing where to locate the date line, and they spent most of the sixth session on Monday, 20 October, hashing it out. Several options were considered. Lewis Rutherfurd proposed 180 degrees from Greenwich. Count Lewenhaupt of Sweden countered with a suggestion that they might follow the example of the Rome Conference, which had proposed Greenwich itself as the date line (meaning the universal day would begin at noon Greenwich time). The Spanish delegates proposed 180 degrees from Rome, which they claimed most of the world already used by way of the Julian calendar (including the Gregorian correction). One Spaniard claimed that a change from this practice would involve a much more complex shift in the calendar itself. Leave Greenwich for navigation, but time and date have always been measured from Rome.

J.C. Adams disliked the Spanish proposal. Reckoning time by one meridian and longitude by another seemed unnecessarily complex. Strachey agreed. Besides, they argued, how could Greenwich cause confusion in the daily use of the calendar? As Strachey reiterated, the universal day had nothing to do with everyday life; it will not interfere in the smallest degree with any purpose for which time is employed in civil life. The two objects are entirely distinct. It is obvious that the conception of the necessity of having a universal day has arisen from the more clear conception of the fact that time on the globe is essentially local; that the time upon any given line (supposing it to be a meridian) is not the time at the same moment on either side of that line, however small the departure from it may be; and for scientific accuracy it has, therefore, been thought desirable to have some absolute standard to which days and hours can be referred.[94]

Strachey wanted his peers to understand that the universal day was a minor reform, really for specialists, and would not affect the public (Fleming, who did want to change civil timekeeping, must have been upset!).

The Spanish diplomat responded that the conference's decisions might have unforeseen consequences, which could spill over into civic life. Who knows "what difficulties we may be driven into? Every country will be obliged to count both ways. They will have to use civil time and universal time."[95] Having multiple times could make things more complicated than was necessary, he claimed. At this moment, Juan Pastorin broke ranks with his countryman and offered the obvious solution – some form of standard time, as·Fleming had been advocating. Unfortunately for Pastorin, the proposal was once again voted down. The debate shifted back to the date line. It soon became clear that some variation on the Greenwich meridian would win out, so the question narrowed to whether the universal day should begin at noon or midnight, Greenwich time.

As for why anyone would propose to start the universal day at noon instead of midnight, it is important to remember that astronomers used a unique system of timekeeping. The astronomical day changed dates at noon, so that astronomers taking observations overnight would not have to change the date in the middle of their work (similarly, sailors often started the nautical day at noon, as the Spanish delegate pointed out).[96] For those who wanted the new universal day for scientific purposes only, aligning it with the astronomical day, not the civil day, made sense. This was the conclusion that the Rome Conference had reached in 1883.

At the IMC, however, a new idea was proposed by J.C. Adams. If railways and telegraph operators were to be using universal time as well, not just astronomers, then perhaps the universal day should align with the civil day. Indeed, the astronomical day could be altered as well, to begin at midnight, despite the inconvenient date change mid-shift. Adams suggested that astronomers could adapt more easily than the public, because they were few in number and intelligent enough to understand the change and implement it.[97]

A few further objections and questions arose, but ultimately the IMC agreed with Adams: the universal day would change at midnight Greenwich time, not at noon. Even Fleming backed the resolution, as it corresponded with the way standard time worked in the United States and Canada. Similarly, a resolution was passed "that the Conference expresses the hope that as soon as may be practicable the astronomical and nautical days will be arranged everywhere to begin at midnight."[98]

After this discussion, the debate moved away from timekeeping. The French delegate put forward a motion for studies on the use of a decimal system in measuring angular space and time. It was a vague proposal that required little effort on anyone's part, and so it passed without much deliberation. Things were coming to a close.

DAY 7

The next session (Wednesday, 22 October) was the last at which any major debates took place. The subject up for discussion was once again the question of standard time. At the end of the last session, Strachey, the delegate for India, had proposed implementing some form of time zones, at least ten minutes, or two and a half longitudinal degrees, wide. Individual nations would decide the exact width they wished to use. However, Strachey now withdrew his proposal, having discussed it in the intervening period with his colleagues and found that there was no consensus on the idea of standard time in any form.

Fleming had spent all of the twenty-first preparing a speech, but Strachey's withdrawal of his motion removed standard time from the agenda, effectively silencing the Canadian.[99] The final word, the summation of the IMC's contributions to the world, was given by Lewis Rutherfurd: "We should not seem, in any way, by our action here, to interfere with the convenience of the world in the use of its present civil time ... Our universal day is for those purposes only for which it may be found convenient, and that it is not to interfere in any way with the use of civil or other standard time where that may be found convenient."[100] It was the end of Fleming's ambitions. Standard time was not to be made a global project.

Behind the scenes, the evenings around sessions six and seven offered several opportunities for informal discussion. On Tuesday, 21 October,

Fleming dined with fellow delegates Evans and Rustem at "the club" (Rustem was Fleming's guest).[101] The next night, at the Metropolitan Club, Rustem returned the favour, inviting Fleming, Evans, and the secretaries of the British legation in Washington.[102] By this time, the real work of the conference was finished, and on the night of Thursday the twenty-third, at the British embassy, the ambassador and his wife hosted a more formal dinner for the four British delegates, Adams, Evans, Fleming, and Strachey. The Russian and Swedish ministers were invited as well, along with their wives.[103]

The next day, Fleming left Washington for good. He did not stay for the final ceremonial end to the conference, which took place on Saturday, 1 November. He was not alone. Many participants left early. President Rodgers wrote to Fleming on 31 October that "so many delegates are gone or are impatient to go."[104] Any appetite for further debate had diminished quickly.

There was a mix of emotions about the end of the conference. Fleming's premature departure suggests some disappointment on his part, but he was ever the optimist, and soon was back to campaigning for time reform (Pastorin aptly named him an "indefatigable propagandist").[105] Fleming would go on to claim (with dubious evidence) that the IMC was a great step on the road to worldwide standard time. Janssen, meanwhile, had failed to prevent Greenwich becoming the prime meridian, but he was proud of the fight he had put up and maintained that France had the moral high ground. Galvan, too, took pride in his efforts to support scientific neutrality in the face of the Anglo coalition, whatever the results.

Adams, in contrast, had achieved exactly what he wanted. He wrote to a colleague; "I am perfectly satisfied with the results of our Congress at Washington in which I took a more prominent role than I expected to do."[106] Adams left the capital on Saturday, 25 October, skipping the final formalities.[107] The career diplomats, of course, stayed in the U.S. capital, carrying on their business. Galvan, now finished with the IMC work, turned again to his trade negotiations with Frelinghuysen.[108] Of those who were not long-term residents in Washington, Cruls, Janssen, Rodgers, and Strachey were among the last to leave.[109]

AFTERMATH

So what did the IMC achieve? In principle, it had established Greenwich as the world's prime meridian, to be used for determining longitude. It had also set up a universal day based on that meridian, but had not prescribed any specific uses for it. It did not establish standard time or time zones in any form. Indeed, its only specific time reform was suggesting that astronomers change their astronomical day to match the civil day. The IMC's impact on civil timekeeping, apparently, was nil.

The immediate and medium-term global response to the IMC resolutions was underwhelming. They became binding only on nations that ratified them. By the end of the 1880s, the only country to do so was Japan, under the influence of Kikuchi Dairoku.[110] Historian Ian Bartky calls this move by Japan "the only tangible result of the International Meridian Conference."[111] France of course refused to use the Greenwich meridian, and would not switch over until the 1910s. Even the host of the conference, the United States, failed to ratify the resolutions. The U.S. presidential election changed the political landscape, and Democrat Grover Cleveland's new administration had little desire to implement its predecessor's plans.[112] Admiral Rodgers kept pressure on the new Congress, but failed to achieve results, telling Fleming, "When the new administration came in, I found no great interest in what its predecessors had begun."[113]

The IMC earned little respect. Its resolutions were treated with such dismissiveness that in 1889 in Paris, at the Fourth International Geographical Congress (a follow-up to the Venice IGC conference in 1881), Italian reformer Tondini de Quarenghi proposed Jerusalem for the prime meridian and called for a new conference to render the IMC decisions obsolete. Fleming, worrying that such a conclave could undermine all his efforts so far to standardize time around the world, pleaded with de Quarenghi to let Greenwich stand.[114] As it happens, the new gathering never occurred, but clearly the IMC was by no means considered definitive.[115]

The IMC results also divided the astronomical community. Many members thought the notion of changing the astronomical day to match the civil day preposterous, and the idea sparked furious controversy. William Christie, the new astronomer royal at Greenwich, intended to comply, announcing that on 1 January 1885 the clocks at Greenwich would be changed to match the civil day, using a twenty-four-hour notation. Some newspapers printed the story with excitement, wondering if the new clock might leak into civic timekeeping as well. "Perhaps in this scientific age it may gradually creep from scientific men and scientific books into ordinary usage," wrote one paper.[116] In the United States, Christie's counterpart, S.R. Franklin, also planned the same shift for 1 January.[117] But not everyone was so enthusiastic.

Simon Newcomb was the most vocal opponent of the change in astronomical day. It "is not merely a change in habit ... ," he wrote, "but a change in the whole literature and teaching of the subject. The existing system permeates all the volumes of ephemerides and observations which fill the library of the astronomer."[118] What is more, the *Nautical Almanac* was prepared several years in advance, so could not make the shift immediately, or for four years it would read incorrectly.

Franklin did not agree with Newcomb's objections. He wrote, "It seems to me eminently proper that the nation which called the Conference should be among the first to adopt its recommendations."[119] Yet the opposition gave him pause, and he soon wrote to other American observatories for their opinions, as well as to Christie in Britain.[120] Only two of the eleven responses agreed with Newcomb, and the rest were open to change.[121] Yet Newcomb's influence was such that Franklin gave in at the last moment, deciding on 31 December to postpone the change, at least until the *Nautical Almanac* could be altered alongside it.[122]

Kikuchi Dairoku, hearing of Newcomb's opposition, wrote to J.C. Adams on 12 December, asking what the United Kingdom would do, including Adams's observatory at Cambridge. Kikuchi had heard that Newcomb's objection had caused Christie to "defer taking any public action though he will adopt the change in the Observatory itself."[123] Indeed, although Christie fully supported the change, he adopted the change merely for internal use at Greenwich.[124] The *Nautical Almanac*, however, would not be altered.

The inertia of astronomers and naval officers was stronger than anyone anticipated. The astronomical and nautical days shifted only decades later in 1925. In Canada, Fleming and the Canadian Institute continued to campaign for unifying the nautical, astronomical, and civil day. Fleming blamed Newcomb for the lack of progress, writing savage attacks in 1895, claiming that Newcomb had been sabotaging international cooperation over time reform, including creation of time zones, for years. According to Fleming, Newcomb, when asked about applying standard time zones beyond the United States, had answered, "We don't care for other nations, can't help them and they can't help us." Indeed, Newcomb thought the entire scheme "a capital plan for the millennium. Too perfect for the present state of humanity. See no more reason for considering Europe in the matter than for considering the inhabitants of the planet Mars."[125] Fleming claimed that "Newcomb stood alone in his antagonism to this scientific reform in which all nations were concerned … Mr. Newcomb has always been at war with the movement to promote the unification of time reckoning throughout the world."[126]

Written after more than a decade of his own unsuccessful lobbying, Fleming's frustrated revisionist history of how Newcomb undermined the

IMC is understandable, but should be taken with a grain of salt. He conflates the issue of standard time (which the IMC never agreed on) with the much smaller question of changing the astronomical day to match the civil day, and thus attributes the IMC's failure entirely to one man.

But, as we have seen, the IMC's shortcomings were not of Newcomb's making. Indeed, it was a "failure" from only a certain perspective. Fleming and Janssen might consider it that, but if you asked Adams, or nearly all of the other astronomers and navigators (and even diplomats) in the room, the IMC achieved exactly what it was supposed to: establish a prime meridian for determining longitude.

CONCLUSIONS

Many layers of conflicting interests gathered in Washington in October 1884. National rivalries played a role, but, on a more fundamental level, conflict at the IMC emerged out of two competing understandings of the conference's purpose. On one side were the few, like Fleming, Abbe, Allen, and Pastorin, who wished to transform the world's civil timekeeping. It was their efforts that brought about the gathering in the first place. Ironically, they were outnumbered at the IMC by other interests: navigators who desired a standard for longitude, and astronomers who sought a tool (the universal day) for use by scientists and specialists.

The latter groups won out. The majority of the IMC delegates had no intention of discussing sweeping civil-time reform (recall that the British delegates, for example, were chosen by the SAD for their opinions on the metric system, not on timekeeping). For these astronomers, the discussion of a prime meridian and universal day was always about specific astronomical and navigational uses. As Russian astronomer Otto Struve observed after reading the conference proceedings in 1885, its decisions would be most valuable to science, navigation, telegraphy, and railway lines; while "ordinary, every-day life, which in its locality is regulated by the Sun, would not immediately be affected by it."[127] Struve added that astronomers, of all scientists, least required the unification of time, because they were experts on time. They dealt with time differences daily, and conversion was easy for them. But they had no intention of forcing the general population to regulate their clocks in a new way.

In the half-century or more following the IMC, time reform was removed from the international sphere and instead carried out primarily by individual nations, piece by piece. The IMC had been ineffective. France, Brazil, and the Ottoman Empire marched to their own tune, leaving the question of civil timekeeping to individual sovereign nations.

Meanwhile, colonization would transplant European national times to subjugated territories by force. After independence almost a century later, many of these former colonies would remake their own national times. It would take decades of piecemeal legislation to establish standard time worldwide, as each nation made its own decisions. And even then, standard time was not perfected: in the twenty-first century it conforms as much to national boundaries as it does to lines of longitude, and there are plenty of exceptions and temporal oddities, not to mention daylight saving time or summer time, which throws the whole system for a loop twice a year in some countries. Fleming's dream of twenty-four geometrically perfect time zones was never realized.

Yet the primacy of the nation-state in the creation of time zones over the long term was not inevitable. Nations were left to their own devices on the subject only because the IMC left the issue of standard time unfinished, and this failure resulted from professional, not national, competition. Few national delegations at the conference were united among themselves. William Allen overstepped the ambitions of the rest of the American delegation, Sandford Fleming was undercut by his British peers, and Juan Pastorin did not agree with the opinions of his fellow Spaniards. Which persons were in the room, their occupation, their background, their vision of the modern world mattered more to the outcome than which nations were in attendance. The IMC was a failure for the designs of civil engineers and businessmen, but it was a victory for astronomers and navigators. Railwaymen in North America had already demonstrated that international cooperation in civil timekeeping was possible, after all. But at the IMC, astronomers held all the cards, and naval and astronomical interests drove the debate, not those of railway corporations. The network of astronomers that dominated the IMC focused narrowly on scientific pursuits like the transit of Venus, or determining longitude for the surveying of colonized territories, or for navigation at sea. It had little interest in regulating time for ordinary people, preferring to allow local time where convenient, and creating universal time for only those few who wanted it. The results of the IMC, when looked at through this frame of reference, are unsurprising.

We come back now to the central question: Why did the system of time standardization set up in the 1880s take the form that it did? Railway engineers, acting on the need to simplify the growing, complex, global network of trains and telegraphs, worked to standardize time. But that campaign became entangled in the scientific quest for longitudinal and astronomical accuracy, as well as in the politics of systems of weights and measures. The interests of an insular astronomical community dominated the prime-meridian debates, edging out those of railway engineers. This left time reform in the long run up to the discretion of national governments, not all of which were keen to regulate time in the same way that North American railways had. The IMC was not destined to be the definitive moment in standard time's implementation.

If not the IMC, then what else? Given France's initial opposition, we can look to the French for an answer. They finally accepted the Greenwich meridian just before the First World War but did so without diminishing Paris's importance, via the power of a new technology: the radio. From the 1920s on, the Eiffel Tower broadcast a time signal instantaneously across incredible distances, making Paris the means by which Greenwich time was disseminated. The advent of radio made universal time much more generally useful, now that it could be easily shared. Aviation would soon adopt it, as would other industries, making it far less esoteric, and more useful to more people. The norms established by North American railways in 1883, by the IMC in 1884, and by radio in the 1920s, each in its own way, all shaped how human beings measure time.

But this leaves a gap of some forty-odd years between the IMC and the radio. During these decades there was a disconnect between methods of timekeeping. Accurate, universal time existed for telegraphy, railway travel, and esoteric scientific tasks, but was difficult to disseminate to the public. Yet less precise local time in all its variations was in common usage for everyone else. The multiplicity of times sowed confusion, frustration, and occasionally humour. Exploring the way people navigated their way through this tangle of competing timekeeping methods around the turn of the century is the focus of the next chapter.

4

"The House That Jack Built":
Selling Time, Constructing Modernities

The IMC was over. The civil time–reform proposals championed by North American railway engineers had been derailed by the astronomical community, which co-opted the conference to suit its own needs. As a result, the IMC had practically no immediate effect on civil timekeeping. Instead, it established a universal day, based on the Greenwich meridian, for use primarily by astronomers and navigators, not by the general public.

But the fact that the new universal day was not meant to be used by everyone was poorly communicated. Engaged readers around the world read about the IMC results in newspapers and wondered how the new universal day might affect their own lives. "What time is it?" seemed a more complicated question than ever. Ordinary people began to have to grapple with questions about who had authority over time and about whose time was the "true" time. The answer was unclear and would remain so for decades. A simmering contest for supremacy developed between a supposedly perfectible and universal scientific time – produced by expert astronomers to be rigorously accurate, requiring some means of distribution to universally synchronize it with the time of other longitudes, and often conflated with the use of twenty-four-hour clocks in place of a.m./p.m. – and other, more practical, if less precise, forms of timekeeping.

This conflict is worth exploring, because it provides a window into how societal norms regarding timekeeping were moulded, pressed between these two camps. The top-down decision-making of diplomacy and science forms only half the story. Those decisions collided with norms of public behaviour rising from the bottom up. When diplomatic decision-making at the IMC ended, the process of implementation began, and it was messy. Official changes in time measurement were everywhere matched by local debate among ordinary people, and their reactions reshaped the temporal landscape, in ways neither Fleming nor Adams could have predicted.

This chapter focuses on the public reaction to scientific time in Great Britain, where the contrasts between official Greenwich time and other, less precise forms of timekeeping were particularly stark. Britons reacted to scientific time primarily in one of two ways. First, most people continued to use the time that was convenient to them. They ignored scientific time altogether, or treated it as a joke. While they were aware of new ways to measure time in scholarly circles, on railways, and at the IMC, such changes had little bearing on their daily lives. We must be careful, after all, not to overstate the IMC's influence: the world did not suddenly devolve into chaos because there were now too many competing times. Accurate Greenwich time was still largely inaccessible to the majority anyways, being expensive and complicated to disseminate. Yet its existence could hardly go unnoticed. Scientific time was front and centre in the cultural milieu, part of the Victorian quest (at least in Britain and its sprawling empire) to quantify and measure the world and hence a symptom of nineteenth-century visions of modernity and progress.

This leads us to the second way that people commonly reacted to scientific time – they made use of it as a status symbol, to appear modern and forward-thinking, to lend credence to their businesses, and to establish themselves as "legitimate" authorities on future developments in human technology and society at large. The universal day promised progress, and those who wished to appear modern attempted to gain access to scientific time, and to claim it as their own. Both the attempt to gain access to universal time and, equally, the business of disseminating that time to others became battlegrounds for competing visions of the future. Just as at the IMC, where nations debated over who should own the prime meridian and professions fought over its uses, so the public turned civic timekeeping into a debate about who should control it. Whose was the "true" time, and who should have access to it? The stakes were high, because those who had access to it could claim to hold the future in their hands.

To summarize, people both satirized scientific time and coveted it. Both of these impulses – the tendency to mock scientific time and the desire to be seen using it, or at least seem knowledgeable about it – arose because scientific time was hard to come by. This is a story of the social ramifications of its limited distribution. Being as restricted as it was, scientific time was not a practical option for everyone. Yet that scarcity, along with the legitimacy lent it by the IMC's imprimatur, made it highly desirable in some circles.

PUBLIC REACTIONS TO THE IMC IN BRITAIN

Before the IMC, Britain already used a dizzying number of unique time-keeping practices. Local time was still measured by the sun or stars in many towns and villages, while railways ran on Greenwich time (Ireland followed Dublin time instead). In Britain, as around the globe, religious institutions were often the most prominent markers of time for communities, as were town halls and other public buildings. Competing local, imperial, and religious times battled for prominence.[1] The IMC did not do away with this multiplicity of times, but added another layer to it. Now there was a universal time, based on Greenwich, which was meant to be authoritative. All other times were supposedly subservient to it. Yet it was cloistered, inaccessible to most of the population. In practice, its existence merely made telling the time more complex. There was nothing universal about it at all.

Many Victorian Britons responded to the confusion by poking fun at it – it became a regular punchline in cartoons and humour columns. As historian Robert Darnton famously suggested, when a modern reader can no longer understand a joke from the past, something significant has changed – there has been a shift in *mentalité* – something about that old society has been lost in the new, rendering the old punchlines unintelligible.[2] That is not the case here. Unexpected time changes and frustratingly incorrect clocks still trouble people today, enough for us to commiserate with these nineteenth-century woes. Yet not, perhaps, to the same extent. A lack of temporal accuracy appears to have been unpleasantly ubiquitous in Victorian Britain.

One humorous story from the *Aberdeen Weekly Journal* in 1887 featured a working-class fellow asking the police about the time:

> He sort of squeezed himself into police headquarters, hat in hand, and shambled up to the desk, bowed very low, and inquired – "Am da boss officer in?"
>
> "Yes, sir."
>
> "Wall, boss, I wants to know 'bout dis time bizness. I've bin havin' a heap o' trubble fur a week past."
>
> "What time are you running on?"
>
> "Dat's what I want to find out. One feller he tells me to go on solar time, an' another tells me standard time, an' my ole woman she's got a third time, an' Ize all mixed up. I told de ole woman dat I was comin' down to get purleece time an' stick to it."
>
> "Well. Set your watch at 1:28."

"Yes, sah. Dat's de fust satisfacksun I've had in two hull weeks."
He pulled out an ancient "turnip," felt around for a key, and had
just got ready to set the hands when the crystal fell out and smashed,
there was a long continued whirring among the works, and as he
held the timepiece to his ear and shook it the internal mechanism fell
on the floor, and rolled under a bench.

"I speckted sunthin' of the sort," said the man as his chin began to
quiver. "Dat comes of tryin' to run on three sorts o' time. No watch
kin stand any such foolin' as dat, an' I might a knowed it."

"What will you do now?"

"Nuffin'. Dat settles time on dis chicken fur de ne' six months, an'
Ize gwine to git up in de mawnin' when Ize hungry, an' go home at
night after the old woman has got de wood in."[3]

The working man reacted to the "scientification" of time by shunning
it altogether, returning to natural rhythms in a knee-jerk rejection to a
modernity that looked more complex than convenient.

In a similar vein, the *Hampshire Telegraph and Sussex Chronicle*
joked in July 1885, "St. Louis [Missouri] has standard time, meridian
time, Southern time, Western time, and so many other kinds of time that
only a crazy man carries a watch."[4] The U.S. railways, with multiple
time zones across the continent, seemed to British observers even more
convoluted than the situation back home. But British papers told tales
of temporal misadventures close to home too, like this one from the
Manchester Courier and Lancashire General Advertiser in 1895:

A Gentleman rode up to a small boy sitting on the fence in front of
his home and inquired if he lived there.

"I try to," was the response.

"Well, my boy, I want to know what time it is; can you tell me?"

"Yes, I kin; I wuz in the house just five minutes ago, and the old
clock was pintin' at eleven."

"What kind of time do you have?"

"Oh, us have all kinds."

"But I mean do you have solar time or standard time?"

"That's what I said. We have all kinds."

"I don't understand you."

"Don't you? Well, come to our house and live a while, and yer'll
learn. My sister Sal she has standard time – that's the clock; I has
city time – that's the town clock; the hired girl has sun time – that's

watching the shadders; and pap and mam has a deuce of a time –
that's what they're doing in there now, and I'm settin' on the fence
till they get her reggerlated. By gosh, you hadn't better wait roun'
jere if you don't want to hear suthin' strike, an' strike mighty
durn hard."

The man rode away rapidly, and the boy kicked another plank off
the fence.[5]

The punchline plays off middle-class stereotypes regarding turbulent
working-class marital relationships and errant, unsupervised children,
but confusion about timekeeping was just as prevalent a trope in late
Victorian society.

The *Manchester Times* humour column in 1889 squeezed in a similar
short segment on time. A bemused inquirer asks: "Stranger: 'what time
is it please?' Scientific man (absently): 'What do you want – sun time,
mean local time, or standard time?'"[6] There is no further punchline –
none was needed.

Of course, confusion and bemusement were not the only reactions
to inconsistent timekeeping. It was also an opportunity to exploit.
Mischievous school children used and abused it, and, equally, were
foiled by it. For example, at Oxford, the clock tower of Christ Church
College, known as Great Tom, struck local time, which was five minutes behind all the other clocks in town, which kept Greenwich time.
According to a story in the *Observatory* from 1908, one student, arriving a few seconds after his 9 p.m. curfew, protested that Great Tom
had not yet stuck. The porter refused him entry, quipping that the curfew rule was centuries older than Great Tom.[7] Other students used the
five-minute gap to argue for early release from classes. As one alumnus
recalled, "We earnestly contended and thought that we ought to begin
by the later and end by the earlier, thus effecting a savings of ten minutes in the hour."[8]

Cartoons about time changes found their way into *Punch* magazine
in 1884, just weeks after the end of the IMC, which had decided to
count universal time up to twenty-four hours instead of twelve. A car-
toon from 13 December (see Figure 4.1) shows a baffled Father Time,
unable to read a new twenty-four-hour clock installed at Lincoln's Inn
in London. The cartoonist suggests putting up a twelve- and a twenty-
four-hour clock side by side, allowing passers-by to "choose their own
time, rather than having the new, 'scientific' 24 hour clock imposed on
them without choice."[9]

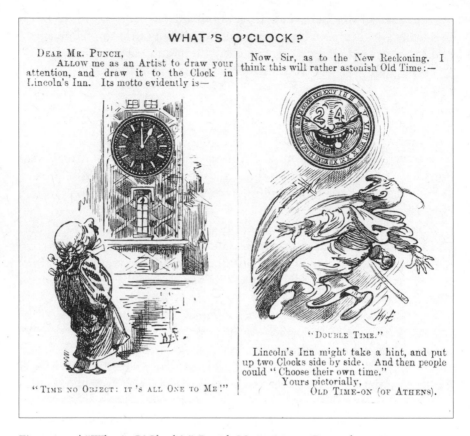

Figure 4.1 | "What's O'Clock?," *Punch Magazine*, 13 December 1884.

A careful reader of these jokes and stories might also notice that telling time "properly" was a marker of status. For example, the servant girl in the *Manchester Courier* column knew how to tell time only by shadows, while her employers used various forms of clock time. The more mechanical the time-telling method, the more modern, the more respectable. Yet scientific time-telling devices that seemed to complicate time-telling, rather than simplify it, alienated people. Scientists like the absent-minded academic from the *Manchester Times* column appear overzealous in their search for accuracy, complicating what should be a simple question. John Rodgers, superintendent of the U.S. Naval Observatory, remarked in 1881 that scientists "sometimes overestimate their functions," destroying the simplicity of everyday life.[10] According to Rodgers, "The people who do not care for scientific time are a thousand for the one of those who do."[11] Such disdain for scientific overreach is captured in another brief story in the *Observatory*, in which Astronomer Royal George Airy made a particularly lengthy speech. Afterwards, a bored Lord Palmerston was said to have whispered under his breath, "Is there not some confusion between Greenwich Time and eternity?"[12] Astronomers were considered brilliant but impractical, unnecessarily complicating the rhythms of daily life. What Britons of all classes wanted was something in between: accuracy and modernity, yes, but also simplicity.

The social implications of timekeeping went beyond the newspaper funny pages. Time and its measurement were also connected to political, economic, and social movements, such as the quest for worker's rights, which was often couched in the language of time. Tom Mann's 1886 pamphlet "What a compulsory eight hour working day means to the workers" helped spark the Eight Hour Movement, in which the Fabians, the Social Democratic Federation, trade unions, and other workers' rights and socialist organizations pushed for legislation to shorten the working day.[13] They framed the well-being of workers not so much through working conditions or wages, but through the measure and use of employees' time. Although the movement failed in the short term to enact laws limiting the workday, it connected time and work in the public consciousness in the 1880s, much as the Factory Acts of the 1840s had done when they enacted the ten-hour workday for women and children. Measuring time, then, had wider political and economic implications, particularly in the charged reformist atmosphere of Britain in the 1840s, and again in the 1880s.

Time, its measurement, and its cultural meaning also found its way into popular literature. Jules Verne's 1873 story *Around the World in 80 Days* is perhaps the most obvious example. The plot relies on a time-keeping error for its climax, in which the hero, who has made a bet to travel around the world in eighty days, believes he has failed and lost the bet by a slim margin. However, because he travelled eastward, he unknowingly gained a day as he crossed the date line, winning the wager after all. In Verne's capable hands, this temporal oddity made for a dramatic change of fortune.

Bram Stoker's classic 1897 gothic horror *Dracula* also employs time and timetables to raise the stakes. The heroes race across Europe, calculating railway timetables and steamship speeds to outwit the undead threat. As literary scholar Adam Barrows writes, "*Dracula* narrates the violent struggle of the last vestige of an 'outside' to standard time's grid." The landscape surrounding Dracula's castle is entirely unmapped, a

Figure 4.2 | Martial Bourdin, the Greenwich bomber.

vestige of an older age. "The ultimate elimination of that vestige, or more accurately, its transformation into a temporally synchronized narrative, provides the fin de siècle foundation myth for an empire of temporal uniformity."[14] Both in literature and in reality, the universal day established by the IMC was symbolic of the drive to modernize the world, to

Figure 4.3 | Onlookers at the scene of the Greenwich bombing attempt.

count and measure it, to make it knowable and erase the unexplored edges of the map. In *Dracula*, scientific time was meant to overcome the unmodern, the antiquated, and the immoral, eradicating the old method of local solar time.

But that process of eradication was more successful in literature than in the real world. Confusion about, and opposition to, scientific time abounded. Perhaps the most explosive example of opposition to the new time changes came in 1894, when a French anarchist named Martial Bourdin (see Figure 4.2) attempted to bomb the Royal Observatory, Greenwich. His bomb exploded prematurely, killing him.[15] Bourdin's motivation to attack the observatory came from the symbolic place Greenwich had gained as the physical centre of the temporal world order. The attack on Greenwich was an attack on British imperial authority itself, represented by its all-important timekeeping apparatus.

A decade and a half later, in 1913, there were rumours that suffragettes might make a similarly symbolic attack on the observatory. A gentleman

told Scotland Yard that he had heard "two well known suffragettes con-
versing on a tram car. He heard them say 'Wait till they start on the
Greenwich Observatory, London. [Living] without time will cause them
to wake up.'"[16] As a result, police increased their presence around the
facility and kept patrols there for more than two years.[17] Violent threats
to Greenwich like these captured the public imagination (Figure 4.3).
Joseph Conrad, for example, fictionalized Bourdin's botched bombing
in *The Secret Agent* (1907). People were interested in reading about the
conflict between order and chaos, and about temporal perfection and its
unattainability. Victorian and Edwardian authors fulfilled those wishes.

Temporal themes in literature did not just reflect changing timekeeping policies; some helped shape them. While preparing for the IMC, for
example, Frederick Barnard came across a novel in which timekeeping
was essential to the plot, and he wrote to Sandford Fleming about it:

> I came across a curious illustration, the other day, of the blunders a
> man may commit who is ignorant or inattentive to the differences
> between local times under different meridians. It occurs in a novel of
> a rather sensational character, but possessing a great deal of liter-
> ary merit. The devil of the story is a married man who deserts his
> wife and runs off (from England to the continent) with a charming
> young woman who supposes him single and whom he promises to
> marry. But he puts her off from year to year till he becomes tired of
> her and jealous. He breaks down in health, and finally conceives a
> diabolical scheme by which to inflict pain and injury upon her. To
> her surprise and delight he proposes at last to marry her, and does so,
> but then makes a will of which he acquaints her with the provisions
> and in which he leaves all that he has "to his beloved wife." His own
> death directly follows. On the very day of the marriage (at Naples)
> the "beloved wife" dies in London, and the poor injured girl finds
> herself without money and without character, or else the heiress to
> a handsome fortune. "It depends on the hour" says the lawyer. If the
> marriage at Naples took place earlier than the death in London, the
> man is a bigamist and the mistress is ruined. If the death in London
> occurred before the marriage in Naples, the erring fair one is reha-
> bilitated and wealthy. On careful investigation, it is found that the
> London wife died just before the public clocks struck the half hour
> after nine. The novelist seems to think that his point will be effective
> in proportion to the narrowness of the margin on which he sums.
> So he causes it to be discerned that the priest in Naples commences
> the marriage service at a quarter before ten. But as half past nine at

London means half past ten in Naples (a fact which the writer forgot or never knew), there was ample time for the completion of the marriage service before the life of the London wife came to an end. So that it was a case of bigamy, in fact, though the writer intended it should not be. It amused me that I should encounter so curious a misconception just as this subject is exacting so lively an interest with us.[18]

Barnard caught a plot hole that other readers, less invested in time reform, might have missed, but the novel raised questions about the correct way to measure time internationally, and used temporal accuracy as the linchpin of the plot. It was exactly these sorts of mistakes, of which Barnard accuses the author, that Fleming's system, if it became universal, was meant to fix. But, as we have seen, it did not. The difference between local, standard, and universal times continued to cause confusion, compounded by the varying reliability of the clocks that measured them.

Confusion about timekeeping was not new to the 1880s. As with the hours-of-work movements, the time debates of the 1880s echoed those of the 1840s. British railways had begun to run on Greenwich time in the late 1840s, sparking the first wave of debates about Greenwich time. An 1848 column in the *Glasgow Herald* urged readers to change their own clocks and watches away from local time to match the new railway time. "All that is needful to ensure the general and cordial adoption of uniform time is to prepare the public mind for the change ... Those who are acquainted with the story of the discontented pendulum will recollect that the farmer found his watch wrong a few minutes one fine morning; and this would be the sum total of inconvenience to the public. Having set their watches with the railway clocks, things would be all right again ... the change would inconvenience astronomers only, and they are quite able to take care of themselves."[19]

The "discontented pendulum" relates to an eponymous short story (first published in 1855) by English poet Jane Taylor (1783–1824), common in schoolbooks. In it, an anthropomorphized clock pendulum, owned by a farmer, refuses to work any longer, overwhelmed by considering its many thousands of swings in the coming weeks, months, and years. After a few minutes, the rest of the clock's components convince it to continue its task, one swing at a time. The farmer wakes to find his clock a few minutes slow. Moral lessons aside, the column writer was pointing out that no one, after making a single, simple time change of a few minutes, would ever notice the difference again. This argument, made in 1848, resembles those of the 1880s. It criticized "half-measures" such as keeping both Greenwich and local time, by adding extra hands to a clock dial, for example – these half-measures merely added to the confusion.[20]

But in the 1840s not everyone was ready to accept the change easily. Exeter, apparently, was one of the first communities in the west of England to adopt Greenwich time.[21] But officials at Exeter Cathedral, whose clock was the most prominent time signal available, refused to change its time.[22] A similar event took place in Accrington. An anonymous letter to the editor of the *Blackburn Standard* wrote, "The inhabitants of Accrington and its neighbours would, I am sure, feel very grateful if the parties in authority would keep the Church clock, (there being no other public clock in town) by Greenwich time; as its being kept so irregular causes a great deal of disappointment and chagrin to the inhabitants. Strangers in particular are very often left by the Railway trains, no fewer than twelve or fifteen persons being so left on Wednesday morning last."[23] Steeple clocks had long communicated the time for their communities in Britain, and timekeeping became an ecclesiastical debate as much as a practical one.

Church opposition to the time change was common in some circles. One column writer derided one Reverend John Cummings, who believed Greenwich time was a papist conspiracy. Cummings preached that "to make Exeter, and Plymouth, and Glasgow all preserve the same time as Greenwich, is just to make them tell lies – unblushing chronological lies – to make the church bells tell lies, ladies' and gentlemen's chronometers to lie – in fact, to enact lying by the law of the land ... It is essential Popish, for it is sacrificing truth to uniformity ... I hope you will keep protestant watches."[24] Not everyone took such extreme positions, but widespread public debate ensued.[25] A more level-headed objection came from rural Wales – a complaint that Carnarvon and Beaumaris were both to switch to Greenwich time, even though neither town was "in the immediate vicinity of a rail-road, and yet the arrangements and convenience of rail-roads are the only pretence that can be assigned for thus disturbing an order of time-keeping which has existed since the computation of time began."[26] This seemed a reasonable objection for a place with no railway travel.

But the railways expanded quickly, and the infrastructure for distributing Greenwich time went with them. Under the direction of George Airy, telegraph wires for the transmission of time signals began to be installed alongside the rails in the 1840s and 1850s. Meanwhile, a clock displayed at London's Great Exhibition in Hyde Park in 1851 was purchased by the Great Northern Railway Company, to be placed at King's Cross station, where it would transmit Greenwich time to other stations up the line once the electric telegraph lines had been completed.[27] The Great Exhibition itself helped cement rail's primacy, as it drew "the greatest increase of passenger traffic on the railways" yet recorded.[28] The exhibition, according to historian Derek Howse, "resulted in travel in Britain on an unprecedented scale," requiring new levels of accurate scheduling.[29]

What this meant was that in Britain, Greenwich time already dominated local time by the 1880s, and had for decades, at least in the major cities. Yet the 1880s saw a revival of conversations about which time ought to be used. Three events brought the subject back into the public eye. First, Greenwich time finally became legal time in Britain in 1880 (and ditto for Dublin time in Ireland). This change affected the closing and opening times of shops, pubs, and public offices, for example. Second, the North American railways switched to time zones based on Greenwich time in November 1883, and, third, the IMC of 1884 introduced a universal day for scientific purposes, using a twenty-four-hour clock. While these last two events little affected civil timekeeping in Britain, they pushed the topic back into public discourse. Furthermore, the IMC's resolutions appeared to make Greenwich the centre of time for the world, in theory. Greenwich time seemed to be the true time, at least officially. Yet gaining reliable access to it remained a challenge, and other forms of timekeeping refused to disappear. Although the IMC could quite reasonably be said to have had no tangible impact on public timekeeping in Britain, it nonetheless helped reopen conversations about time and authority, about modernity and accuracy, conversations that had been dormant since the 1840s.[30]

News of the IMC was printed widely in British newspapers. The conference's bitter British-French antagonism attracted some papers, while others merely summarized its resolutions. Only a few papers went into more detail, some predicting momentous changes in timekeeping.[31] Speculative articles pondered when and if scientific time would filter into common use.[32]

Outside of newsprint, clubs and institutions also ruminated on the consequences of the IMC's resolutions. Scientific bodies like the Balloon Society in London discussed them in detail at their meetings, for example.[33] But one of the first groups to learn about the IMC and its possible ramifications was children. While timekeeping was not on the British school curriculum, it was taught informally. Picture books taught how to tell time alongside moral lessons. And in response to the IMC, two evening lectures for "juveniles" were held in London. The Society of Arts put on the first on 31 December 1884, and the second a week later on 7 January 1885. Astronomer Norman Lockyer, inspired by the IMC's resolutions, titled the series "Universal Time: Our Future Clocks and Watches."[34] Both events sold out.[35]

Lockyer began his first lecture with a whimsical flourish. "Once upon a time, ages ago, when the world was very much younger than it is now, and there were very many more elves and fairies than there are now, and even long before Santa Claus was born, and was going about as we hope she [*sic*] will be going about so merrily tonight – all that while ago I say you can quite understand that people had not any clocks and watches."[36] Having caught the children's attention with Santa Claus and elves, Lockyer described the history of timekeeping, and the internal workings of a modern clock, including how barometer pressure and temperature make clocks run slow or fast. Then he arrived at the heart of his presentation – the IMC and the possible changes it might bring to timekeeping. "Now the reason I suppose that I am talking to you now, on New Year's eve – when it is too bad of the Society of Arts to have anybody lecturing or being lectured to – is that on this particular New Year's eve a very wonderful thing is going to happen in connection with time, which will be remembered down the centuries. At midnight tonight, one of the assistants at Greenwich Observatory will go and put back that wonderful clock, which I hope many of you have seen, showing the astronomical time at Greenwich."[37]

Lockyer was of course referring to the IMC resolution that recommended changing astronomical time to match civil time, beginning at midnight instead of noon. This change, as we saw in chapter 3, was rejected at the last minute in the United States, and in Britain the Greenwich Observatory made the change, but only internally, rather than adopting it officially. It would not be until 1925 that astronomers made the shift that Lockyer was describing as imminent. But the astronomer was caught unawares in late December 1884, not knowing that the plans put in motion at the IMC would fall through. So he continued his presentation confidently as if the change in astronomical time was going to occur just a few hours after his lecture.

Lockyer rambled a bit after that point. He left time reform to teach the basics of physical geography to his young audience. He explained that the world was round, and large, and spinning, and why people did not fall off the globe and float out into space. He wanted to establish this background firmly before diving into the complexities of global timekeeping. But time ran short, or the youthful audience grew restless, so the conclusion had to wait.

Lockyer had a lot of ground to cover, and the next week he wasted no time delving into it, with no talk of elves or fairies this time. He described how astronomers use a transit telescope to measure exactly the length of a day, as the spinning earth caused a chosen star to pass across the instrument repeatedly. He then pointed out that different countries did the same task in their own observatories, leading to differences in both nautical maps and timekeeping. Although local time was useful, Lockyer argued that "one set of maps ought to do for all the people in the world."[38] A single prime meridian would make that possible. He summed up: "There is use for time beyond the uses of everyday life, and there are uses for longitude besides the absolute necessity of knowing how many miles it is from place to place. These difficulties and others like them have been growing for years, until at length, last year, there was a meeting of wise men at Washington, and I am going to conclude my lectures by referring to the conclusions at which these wise men arrived."[39]

Lockyer then listed the major resolutions of the IMC, and stated that the new universal day, based on Greenwich, would be counted up to twenty-four hours instead of twelve. "It is that which is going to alter all our clocks and watches," he predicted. "Some people say, 'Oh, this will not come in our time. It is a thing which is all very well for astronomers, people who look at comets and such outlandish things, but we shall not want it.' But I think you will want it, for the reason that it is so very convenient."[40] Telegraph companies, he explained, will want to use it to standardize their practices around the globe, as will the railways with which the telegraphs are "closely associated."[41] Here Lockyer parroted Fleming's refrain that using twenty-four-hour time would end any confusion between a.m. and p.m. He then showed various examples of patented new clockface designs, each showing twenty-four hours rather than twelve. He also mentioned that clock bells would continue to strike only twelve, despite the predicted change in clockfaces.

In concluding, Lockyer assured his young listeners that universal time, and its twenty-four-hour notation, would be meant not for astronomers alone, but for everyone. "If the railway companies and the telegraph companies adopt this time, we shall all of us have to do it."[42] This sentence is representative of the ways the IMC resolutions could be misconstrued to and by the public. Lockyer, and many like him, assumed that the IMC changes would indeed affect civil timekeeping, and he was preparing the next generation for it, even though the conference stated that universal time would not affect civil time. Like the newspaper cartoonists, Lockyer was preparing for a revolution in time measurement, but he was hopeful about it, while they were hesitant and confused. But in both cases, the IMC had put reform of civil timekeeping firmly back into public discourse.

Not everyone agreed with Lockyer – certainly not the majority of professional astronomers, as we saw at the IMC. Although his lectures were aimed at children, adult critics took pains to correct him. The *Horological Journal*, a publication for British watch- and clockmakers, presented several criticisms. Some were minor – making fun of his misconceptions about how a clock works (in particular, Lockyer apparently misguided his young audience as to the sound a watch makes when the balance spring breaks).[43] But one reviewer rejected his thesis that ordinary people would soon have to adopt universal time. Edmund Beckett was a prominent designer who had constructed the clock mechanism that controlled Big Ben at Westminster.[44] He thought Lockyer's lecture material "suitable enough" for juveniles, "but when he got beyond the appropriate elementary information into schemes for futurity, he appears to have … imputed to the astronomers at the Prime Meridian Conference sundry things of which they were not guilty, according to all the authentic reports that have been published. I repeat that they never said one word about an universal civil time."[45] Beckett pointed out that few delegates intended to alter how the public measured time. He rejected too Lockyer's assertion that telegraph companies would adopt universal time, and that this would force the public to change as well.[46]

Most members of the clockmaking industry fell in between Lockyer's confidence and Beckett's scepticism. Some cited the influx of new patents for twenty-four-hour watch dials as evidence of a coming change.[47] Others were not so sure. In the months after the IMC, the pages of the *Horological Journal* were chock-full of arguments from both sides, and hints that people were betting on one or the other, some ready to cash in with patents for new watch dials.

These entrepreneurial endeavours perturbed one contributor. The influx of watch-dial patents coincided with a broader public debate about patent laws and free trade, and how much intellectual property should be protected. The year 1884 saw the most British patent applications ever, due to recent changes in the application process.[48] The contributor felt that it was almost too easy now, and "it would be an intolerable interference with trade if every little simple device that would suggest itself to the minds of most men after a few minutes' thought were be allowed to be patented … for instance, several people have applied for a patent to secure the placing of a second hour circle, with numbers from 13 to 24, on watch dials."[49] But the writer was not against twenty-four-hour time itself. He or she simply opposed the idea that a few quick-thinking clockmakers would dominate the market and reap all the profit from time reform, because they were the first in the patent-office door. Instead, all clockmakers should have an equal shot at implementing universal time in watches and clocks for the public. "Some few horologists," the missive continued, "speak contemptuously of the proposed universal time as simply a passing craze, but the inconveniences and perplexities arising from the use of local time in foreign communications are so great, and are becoming so increasingly apparent, that the establishment of some such system by civilized countries cannot be far off; and it ill becomes Englishmen to throw cold water on the project," since the IMC's chosen prime meridian lay within Britain.[50]

The *Horological Journal* was full of contributors who, like Lockyer, believed a change to twenty-four-hour universal time was inevitable. The journal had been keeping track of the North American time-reform movement and had published a description of U.S. railways' adoption of time zones in 1883.[51] It followed the IMC closely as well, writing in November 1884 that the conclave had established a universal day and also showed "indications that the ridiculous custom of dividing the civil day into two periods of twelve hours each … will shortly give place to the more natural method of treating the day as a whole … The change must come, but the public require leading a little."[52] Clockmakers, the author argued, must take that lead, and create watches for the new system.

In the December issue, a watchmaker named Thomas Wright opined on the impact of the IMC's resolutions on the clock trade. He was unsure whether public timekeeping would change immediately, but seemed confident that eventually a universal twenty-four-hour time would become the norm. In response, he suggested that English clockmakers should begin creating new mechanisms for clock striking.[53] A Swiss reader replied, offering another way to accommodate twenty-four hours on watch faces.[54]

In January, journal contributors also pondered whether church bells should strike up to twenty-four hours, or other possible ways to signal the new dispensation.[55] Another writer noted that clocks striking up to twenty-four in the late evening might "occasion a considerable disturbance to nervous or sickly persons."[56] These annoyances were overcome in Italy, which already used twenty-four-hour time, by dividing the day into four parts of six hours each, the author explained – Britain might do the same.[57] Indeed, in early 1885 concerns about the striking of public clocks at night, led to one clock's chime in Hastings being silenced between 11 p.m. and 5 a.m., "on the ground that they are a public nuisance."[58]

In the January 1885 issue, J. Haswell summed up the general feeling among watchmakers about whether the public would use universal time. Its adoption, he said, was "more or less probable."[59] Another column agreed: "The determination of astronomers to begin the day at midnight will materially hasten the general adoption of the rational style of reckoning. Officials of the leading railway companies have already been making inquiries with a view to the introduction of the system."[60]

The vicar of one London church, St Mary Magdalene's in Munster Square, took the initiative to make the change himself. He used twenty-four-hour notation to list the Christmas services, quipping to his apparently frequently absent congregation that, "as a little reflection will be necessary to make them out, you are sure to remember them."[61]

The furor over the imminent time change calmed down a little in March 1885, when the *Horological Journal* finally learned that Astronomer Royal William Christie had acted prematurely by changing the clocks at Greenwich to match civil time, and that most other astronomers had refused to follow suit. Christie was apparently "called to account for his precipitate action in the matter," presumably by the observatory's supervisory board of visitors.[62] Edmund Beckett, who had criticized Norman Lockyer's 31 December prediction of imminent changes to civil time, now insisted that changing the civil day to match the twenty-four-hour universal day would require an act of Parliament.[63] Beckett was right in principle, though not necessarily in practice. For example, most of Britain's cities had been using Greenwich time for about thirty years before it became legal time in 1880. Norms of behaviour concerning time tended to precede legislation, not follow it. The same might occur in this case: if twenty-four-hour clocks became the norm, the law would probably catch up, sooner or later.

Beckett's editorial, along with news that Christie had acted out of step with his fellow astronomers, did not deter all watch- and clockmakers from expecting imminent change. One prominent firm, Kendal and Dent, keen to show its "practical belief in the new order of things," offered a £100 prize for an essay contest administered by the Balloon Society.[64] The topic, taking its lead from Lockyer's lectures, was "Universal time, or our future Watches and Clocks."[65] The same company had recently put out a design for a watch dial that showed both twelve and twenty-four hours, seeking to convert people confused by the extra hours on the dial.[66]

The *Horological Journal*'s editors took a neutral stance on universal twenty-four-hour time, but were clearly aware of its significance to their industry: "Whichever way the matter may eventuate, the Patent offices will be the richer for the twenty four hour proposal."[67] They explained that over ninety applications had been received by the patent office in Britain on the subject, and three times as many in the United States. If a change was going to happen, money was to be made. Everyone wanted to "secure the monopoly of what each one believes to be the only true solution to the problem."[68]

The potential monetary implications made the debates over twenty-four-hour time particularly vitriolic within the British watch and clock trade. Firms feared failing in the face of cheaper, mass-producing competitors and of innovations by foreign watchmakers.[69] A global economic depression had begun in 1873, and Britain felt its effects through to the 1890s. Worried clockmakers carefully scrutinized Swiss and U.S. competition. In March 1886, the British Horological Institute and the City and Guilds Institute held a joint meeting in London to consider "the cause of the present depression of the watch trade, and how far it has been brought about by the hallmarking of foreign watch cases."[70] Members of Parliament were invited, in hopes they might assist – whether through free trade or its opposite, protection. The institutes blamed their troubles on an influx of foreign watches, which they claimed (probably unfairly) were "imitations" of their British counterparts.[71]

But other observers looked closer to home for the cause of the depression in the watch trade. Some suggested that British clockmakers were refusing to modernize. "Times have changed, and the tactics of our grandfathers no longer avail. Those who rely on them may expect a gradual but nonetheless sure extinction of their trade."[72] In London's Clerkenwell district, in Islington, where many of the country's best watchmakers worked, the mood was gloomy. One observer wrote that "manufacturers here hesitate to put sufficient capital into their businesses."[73] Their products still used "obsolete" key winders, because watchmakers refused to upgrade production.[74] Clerkenwell tradesmen, they claimed, were to blame for their own failures. More observant contributors realized that the depression was not localized, pointing out that American watchmakers were facing similarly dire times.[75]

In either case, the outlook was grim, so clockmakers grasped at whatever they could to obtain an edge, which explains the high number of patent applications after the IMC.[76] The shift towards twenty-four-hour time, should it occur, promised to end the depression for the industry, as Britons retooled their clockfaces and watch dials. Watchmakers were understandably entranced by the supposed panacea for their economic woes, and their journal continued to monitor the situation carefully.[77]

But the watch industry was not the only one that harnessed time reform for economic gain. New publications and tools were developed to help both specialists and ordinary people sift through the confusion over universal time. Some were sensationalist, making grandiose claims about solving complex problems. One publication, *Vo Key's Royal Pocket Index Key to Universal Time*, proclaimed "The Greatest Discovery Ever Made on Time: Universal Time."[78] That "discovery" was the idea that each watch face could be imagined as a flattened representation of a globe, with the hours marking various meridians. The pamphlet provided tables and charts so readers could use their watch to determine the time at any major city worldwide.

Vo Key's index was far from the only pamphlet of its type. New ephemera were tailored to specific tasks, such as the *Universal Lamp Time Chart*, which helped determine when to light or extinguish streetlamps or vehicle lights, depending on their longitude and time of year.[79] In some cases, timetables were packaged within broader reference materials. *Martin's Tables, or, One Language in Commerce*, for example, included a chapter on standard time alongside an explanation of the metric system of weights and measures.[80] Similar tools were available across the spectrum, from almanacs to railway timetables. Their publication shows a desire to cash in on the public's misconceptions and confusion surrounding the IMC, universal time and its twenty-four-hour notation, and standard time abroad.

It is not entirely clear how well these innovations were received, or who actually purchased them. Most discussions about universal time for civil use remained abstract – people considered it a possible future innovation rather than a new norm. The IMC's creation of universal time based on the Greenwich meridian was never supposed to affect the public, and in large part it did not. But that did not prevent entrepreneurial visionaries, futurists, watchmakers, and engineers from trying to persuade them. So what prevented universal time from reaching a wider audience? The answer is largely technological. Distributing accurate Greenwich time to the public before the radio was expensive and difficult. These challenges contributed to the increasing multiplicity of timekeeping methods in use and also introduced opportunities for innovation and competition in a timekeeping "marketplace."

SELLING THE TIME

If you wanted to know Greenwich time in Victorian Britain, then your best chance would be to check the clock at the local train station. Telegraph lines were the main method of time distribution, and these commonly ran parallel to rail lines. Thus railway stations put city and small-town residents at an advantage. Determined rural dwellers, with some money and knowhow, could purchase a small transit instrument and, on a clear day, make their own observations to determine local time, converting it to Greenwich time with an almanac. But this method was out of reach for most people, both because of its expense and because of the time required to perform it (not to mention the British weather).

By the 1880s, the national time-distribution service had existed for several decades, arranged by Astronomer Royal George Airy and administered by the Post Office. This system connected the main clock at Greenwich Observatory via wire to a receiver clock at the Post Office headquarters. The main clock was manually corrected (just before 10 a.m. and 1 p.m., when the time signals were sent out by the Post Office) by observatory staff – like Annie Russell – who made transit observations with a telescope aligned directly along the meridian. The receiver clock would thus show the same time as Greenwich time, within a fraction of a second (as close as technology and human error allowed), and the Post Office would use that to send out a time signal, via telegraph, to other offices and railway stations around the country. Big Ben at Westminster was also connected to the observatory's main clock, as was a public clock on the outer wall of the Royal Observatory in Greenwich Park. Besides the twice-daily signal, an hourly time service was also available, but only within London itself. Sending the time signal immediately after the main clock had been corrected ensured the least possible margin for error.[81] Even still, the signal was sent by hand, meaning that, while reasonably reliable, it was not perfect.[82]

Of course, the Post Office's time service was only a secondary concern for the Royal Observatory at Greenwich. Its primary purpose was to provide the Admiralty with accurate time signals and chronometers (accurate, expensive watches) with which to measure that time at sea. For this purpose, a time ball had been constructed on the observatory roof in 1833, which dropped at 1 p.m. precisely each day in full view of ships on the Thames River. Members of the public, if in the area, could also see the time ball, but this was a fairly limited audience. In addition, the observatory held regular contests known as the Greenwich trials, in which British clockmakers competed to make the most accurate chronometers. The prizes gave prestige to the top-ranked clockmakers, while ensuring the best possible timepieces for the Royal Navy's ships.[83]

Since the observatory was busy with naval timekeeping, the civil-time service was organized primarily by the Post Office. When, in the 1870s, it began exploring ways to obtain revenue from selling the time, there was some debate about possible types of customers and what they might be willing to pay. The engineer-in-chief wrote that as "applications for time signals are becoming more frequent every day," the Post Office ought to set a regular pricing system, as it had done with companies ordering private telegrams.[84] He suggested making the price prohibitively high, so that only large companies could afford them. Smaller businessmen and "shopkeepers who wish either to save the expense of a transit telescope or simply to advertise themselves" were a pain.[85] They would grumble at every small increase in cost. "We find in practise that these men constantly haggle as to price, and it is this unnecessary haggling or bargaining that appears to me so undesirable for the Post Office."[86] However, the same engineer did not believe that the time service would bring in much revenue and agreed with George Airy that the time should be displayed publicly for free in the window of every major post office, rather than sold, thus avoiding the trouble of processing all these applications for private time signals: "If the signal were thus exhibited at the post office, those who wish to have the luxury of the private signal at their own houses could not object to pay these charges."[87] Despite these initial deliberations, a free time signal at each post office was not put in place immediately, leaving Greenwich time for most of the decade a luxury for those few who could afford it.[88]

By the 1880s, the demand for the Post Office time signal had grown significantly, and a few outlets offered free public time signals, including the one in Cambridge. Each morning, crowds of people gathered outside the facility to hear the time signal called out when it was received on the wire. By May 1881, the daily crowd had grown so large that it was a nuisance, "interfering with the business of the office,"[89] so the clerk discontinued the practice. One regular visitor, Rev. J.B. Pearson, wrote to the Post Office in May 1881 to complain about the end of the service. He owned a chronometer for astronomical observations and was allowed once every two weeks to have it corrected at the Cambridge Observatory. However, it could still not maintain the correct time, "on account of the irregularities even a good chronometer is liable to from the variations of temperature."[90] The post office thus allowed him to correct it more often.

The postmaster general, on inquiring to his staff, was told that Pearson "makes no commercial use of such information, and ... makes a liberal use of the telegraphs. Perhaps he might be obliged with what gives us no trouble."[91] Unfortunately, the official was unwilling to offer special treatment. His priority was not to distribute the time to the public, but rather to eliminate the crowds blocking paying customers from sending telegrams, packages, and letters. Pearson was refused and was told if a daily time signal was that important to him, he could pay to rent a wire directly to his home.[92] This private wire was expensive, so it was also suggested that the university might pay the fee for him, and have the time sent to a university building he could access. Such a service would cost £14 a year for buildings within half a mile of the post office (the price varied somewhat with distance).[93]

Rev. Pearson was not able to pay himself, and at the university both the Philosophical Society, of which he was treasurer, and the Cavendish Laboratory turned him down.[94] But he was not ready to give up so easily. In July, he wrote back to the Post Office reframing his proposition in the language of the public good. Once a week, he argued, the time should be called out to the public, as before, but across the country in every post office, as a regular service for everyone. Pearson's proposal, of course, failed to answer the postmaster general's concern about crowds. But Pearson believed access to Greenwich time was a public service worth providing: "I think that this, if generally done, not only here, but in places where there are first class watchmakers … would be of much service."[95] Pearson went on to explain that, even though many post offices had public clocks outside their buildings, set to Greenwich time, these seldom had "second-fingers," leaving a want of accuracy for astronomers and watchmakers.[96] Time accurate to the minute might be good enough for most of the public, but a weekly announcement of the exact time would be useful to those who required more precision.

What Pearson failed to understand is that to civil servants like the astronomer royal and the postmaster general, access to accurate Greenwich time was not a public right. Universal time was for use by professional astronomers, railwaymen, telegraphers, and navigators, not for ordinary civilians. Anyone outside those categories was not entitled to receive it without a fee. Time, at least authoritative, scientific time, was a privilege.

The postmaster general explained to Pearson that even if the time was called out only one day a week, the crowds would "be still experienced on that one day, and experienced probably to an intensified degree. Not only so, but the Department would soon have to forego the revenue which it derives from Watchmakers who now pay annually for a current sent direct from Greenwich; and I cannot see my way to recommend that this revenue should be given up."[97] The letter explained that the Post Office signal was done by hand and was therefore less accurate than the signal that could be purchased, which arrived direct from Greenwich Observatory itself by wire.[98] Pearson at last conceded the point, writing that if he was in future to carry out any astronomical observations that required such precision, he would pay the fee for the private time signal. At the moment, however, "the extra accuracy secured would hardly be equivalent to the expense."[99]

Pearson was not the only one complaining about public access to Greenwich time in Cambridge. Horace Darwin, son of naturalist Charles Darwin, owned a local company that manufactured scientific instruments. He, like Pearson, had relied on the time called out at the post office to set his instruments. He wrote to the Post Office at the end of 1881 to suggest an electric solution – namely, placing a "simple needle galvanometer" outside every post office, and that the ten-o'clock time signal be passed through it each day, in full view of the public.[100] Darwin reasoned that the time signal sent out daily across the nation, with its thousands of miles of cable, and the time wasted by clerks to pass it on, must be "of considerable expense to the country," and yet so few people had access to it.[101] Installing public signals would make the expense worth the cost. Darwin's need for Greenwich time was self-serving, but, like Pearson, he framed the issue in terms of civic utility.

The Post Office was unreceptive. It could not supply every city and town with this service because of the cost. Nor could it offer Cambridge special access, because then all towns would want it. Like Pearson, Darwin was told that he could pay the fee, like everyone else, to have the time signal sent directly to his home or business. Internally, the postmaster general did consider Darwin's idea, asking his engineer-in-chief to estimate the cost of implementing these electric signals in the major towns across Britain.[102] The figures came in at £7,410 for installation and then £1,065 each year for maintenance.[103] The Post Office was making approximately £1,400 a year from renters buying private or corporate time signals. Providing a free time service would probably mean losing most of those paying customers, as well as incurring an additional £1,065 a year in maintenance expenses.[104] It was not a smart financial decision.

These requests from Darwin and Pearson show that there was a demand for an accurate public time service. There were enough people in Cambridge who wanted access to Greenwich time to clog the post office each day. It is impossible to say if demand was just as fervent elsewhere, but at the very least there was heavy demand for public time signals in areas that housed heavy concentrations of scientific professionals, like the university town of Cambridge, or places with a significant clockmaking trade, such as Clerkenwell in London or Coventry in Warwickshire. Greenwich time as a scientific tool was available to these communities, but for a price. Facing a hefty paywall, relatively few people could afford it. There was now a hierarchy of times. Accurate universal time, restricted by cost, was available to only a few. Greenwich time available for free was accurate only to the minute, unless you happened to live or work close enough to a see a time ball or a time gun (like Charles Piazzi Smyth's setup in Edinburgh) or hear the chimes of Big Ben. The British time service followed IMC prescriptions: universal time was for specialists, and civil time was not to be updated to match the new standard.

Astronomer Royal William Christie, the ultimate authoritative source for Greenwich time, believed that this was exactly as it should be. The public time service was, to him, a luxury, and a distraction from the observatory's real work. In the summer of 1888, the debate over whether Greenwich time was an essential public service or a disposable luxury came to the fore in what was essentially a labour dispute.

The work done at the observatory was outpacing the number of staff available to carry it out.[105] When Christie asked the Admiralty Board for money to hire new staff, his request was denied, and he was told to put his house in order (financially).[106] In response, he threatened to cut any and all "extraneous work, such as the supply of time-signals."[107] "It appears to me," he wrote, "that a condition of [the time-signal's] maintenance must be that arrangements shall be made to enable the proper work of the observatory to be carried on and suitably developed."[108] In other words, Christie was holding Greenwich time hostage until the Admiralty would pay for more observatory personnel.

Postmaster General Henry Cecil Raikes was not immediately aware of Christie's threat, and it is fairly clear, given the evidence, that Christie's goal was not actually to destroy the Post Office's time service, but rather to use it as leverage to increase his budget. It was a bluff. In the meantime, Christie tried to find other ways to cut costs. The previous year, he had asked the Post Office to take over the maintenance of the wires connecting Greenwich to the Post Office, which were old and in need of repair – work that would cost about £150.[109] He had received no answer so in the spring of 1888 raised the issue again. It was a fairly reasonable request. After all, unlike in the United States, where observatories made income from their time service, in Britain all such revenue went to the Post Office. Thus it made sense for the Post Office to maintain the wires. The Treasury agreed, and all parties seemed happy. That was when Raikes learned of Christie's threat to end the time service. He was baffled.

William Preece, the Post Office engineer who was about to supervise the repairs, wrote to Christie: "The PMG [postmaster general] has been frightened by your [decision] ... and won't allow me to proceed with the work as long as the supply of time signals is in question ... What is to be done?"[110] Christie responded that he was waiting on the Treasury: the ball was in its court.[111]

Politicking and subterfuge followed. Preece, ostensibly merely an engineer, became the unofficial messenger between the astronomer royal and the postmaster general. On 9 June, Christie wrote a private note to Preece, explaining why the time service had to be cut, blaming the Admiralty for not providing adequate funding to the observatory's time-service staff. He also somewhat eased his threat, suggesting cutting London's hourly time signals and the national 10 a.m. signal, but keeping the 1 p.m. signal in use for everyone.[112] The next day, Preece asked Christie for permission to show the private letter to Raikes, so that it might convince him to put pressure on the Admiralty to pay up.[113] In other words, this was no longer an attack on the Post Office by the observatory; both branches of the civil service were teaming up to press the Admiralty Board and the Treasury. Upon seeing Preece's note, Raikes, relieved that Christie intended to challenge the Admiralty, not ruin the Post Office, agreed to help. He asked for an official copy of Christie's complaints to use in his own letter to the Admiralty. Christie, pleased to have Raikes as an ally, wrote up an official government memo for Raikes, repeating the complaints he had previously shared with Preece privately.[114]

Raikes performed his role marvellously, acting as if he was shocked by Christie's actions. To some extent, Raikes's frustration was probably real. The two men were uneasy allies. But Raikes wanted to secure the future of the Post Office's time service, and helping Christie acquire adequate funding from the Treasury was the best way to do that. So Raikes wrote to the secretary of the Admiralty in protest: "I think it is imperative to call the attention of the Secretary at once to the serious nature of the proposal of the Astronomer Royal."[115] If the Post Office could not obtain the time from Greenwich, it would have to seek it from the next-closest observatory, Kew. But that seemed absurd. "For what purpose was Greenwich Observatory established, if it was not for the production of accurate time for national and imperial objects, and what object is of more consequence to the Government than the distribution of accurate time throughout the three kingdoms to every post office and railway station! It appears to me that if the Astronomer Royal has failed to convince the Treasury of the need for more assistance, or more financial support he should knock off some other work of less national consequence than the proper distribution of time."[116] Raikes ended with the hope that the Admiralty would press the Treasury to give Christie what he wanted.

After hearing no response, Raikes wrote to the Admiralty again in late July. He began with a history of the time service. The Electric and International Telegraph Company, formed in 1846, had commenced it under George Airy, in agreement with the South Eastern Railway Company. When the Post Office later acquired the sole rights to telegraphs, it inherited Electric and International Telegraph's time-service contracts. Repeating his earlier letter, Raikes then explained the vital importance of the time service both to shipping and to civil timekeeping, in hopes that the Admiralty would ensure its continuance.[117] He also explained that cutting the 10 a.m. signal would generate a chorus of complaints from customers having to switch to the 1 p.m. signal, as it was far more expensive, at £27 per year. (Telegraph lines were less busy at 10 a.m. than at peak time 1 p.m., increasing the opportunity cost of running the latter. The Post Office built that cost into its higher 1 p.m. price.) Raikes concluded by appealing to Greenwich's growing symbolic role after the IMC. As one of his aides put it: "Of all the arguments which can be adduced for the maintenance of Greenwich Observatory, I cannot help thinking that the one which appeals most directly to the popular mind is that the correct time of day is there ascertained and made known."[118]

Raikes again received no response, but his missive had the desired effect – drawing the Admiralty's attention to the time signal's importance. It wrote to Christie in early August demanding an explanation. Why, it asked, was he going to "discontinue a service that was established by your predecessor in the interests of the Public, and especially of the Shipping interests of England, and has been continued for so long a period. Their lordships ... will be very unwilling to sanction the abridgement of a system that is so eminently calculated to improve navigation, and by means of which the chronometers of Her Majesty's Ships are now principally rated."[119] (Raikes had pointed out that watchmakers who made chronometers for the Royal Navy used the time signals to set their instruments.)

Christie offered up a lengthy reply. He would feel perfectly happy to continue the time service, if the Treasury would only supply the funds. He then suggested that the 1 p.m. signal would suffice for Britain's shipping interests. Time balls, time guns, and chronometer rating could be done just as well at 1 p.m. as at 10 a.m., despite what the Post Office said. "This appears to be a question of Post Office revenue rather than of the interests of the public."[120]

Although Raikes and Christie were collaborating to press the Treasury to pay up, neither hesitated to attack the other if it served his cause. In mid-September, Raikes, still waiting to hear from the Admiralty, wrote to it again to point out that the time signal had failed several times in recent weeks. The line and apparatus were in dire need of repair, and the

Post Office had already agreed months earlier to take on that cost. But it would not do so while the future of the time service itself remained in question.[121] This backfired.

At long last, on 1 October, the Admiralty had had enough. It sent two letters. One informed Raikes that the time service would continue, no matter what Christie said, so he should go ahead and fix the apparatus.[122] The other told Christie it would not allow the signal to be stopped. "The issue of such a signal from the Royal Observatory is a duty in direct correlation with the objects for which the observatory was established."[123] It rejected Christie's financial claims, pointing out that both it and the Post Office were working to improve the timekeeping apparatus and hoped that this would reduce time-service labour.[124]

An aggrieved Christie, unwilling to give up, responded on 12 October. He went into detail about the apparatus – operating the 10 a.m. signal was a lot of effort: "The Mean Solar Clock which is used for sending out the time signals is necessarily a complicated and delicate piece of mechanism and requires to be corrected by means of astronomical observations immediately before an accurate signal is sent."[125] It was corrected just before 10 a.m. and 1 p.m. daily (except Sunday, when only the 1 p.m. was corrected). Christie described how employees compared the clock to the previous night's transit-instrument observations and required exceptional skill to make the corrections perfectly in such a hurry. Correcting the clock itself (only one part of the process) could take from ten to twenty minutes, and it was easy to make a mistake.[126] Christie then explained that the 1 p.m. signal was more accurate, with more time to ensure no mistakes or to compare it to other clocks in the observatory in the event of cloudy weather. Almost all naval time signals around the country used 1 p.m., as it was more accurate.

In summation Christie made two things clear. First, "the 1p.m. signal is the only one which it is admirable to use for navigation or other purposes where the greater accuracy and certainty are required."[127] Second, "the 10a.m. signal is a subsidiary signal liable to error on weekdays and not available on Sundays. It is doubtless convenient for the commercial distribution of time by the Post office, but it is unsuitable for purposes of navigation or for the rating of chronometers."[128] Therefore, he concluded, only the 1 p.m. signal fell within "the objects for which the observatory was established."[129]

The Admiralty was not happy with his insubordination. It replied on 1 December, writing that instead of discontinuing the 10 a.m. signal, he should try to improve it.[130] It also corrected his assertion that most naval time signals used the 1 p.m. signal. In fact, most time signals around the country, though fired at 1 p.m., were set using the 10 a.m. signal. Most chronometer-makers also relied on the latter, which was less expensive. Thus the poor quality of the 10 a.m. signal was of real concern. "My Lords regret that you are obliged to speak in regard to the accuracy of the 10AM signal in such a disparaging tone ... Their Lordships would certainly be glad to hear that this Signal could be made more trust-worthy."[131] They wanted to improve it without adding to the labour of the staff, asking if a new, separate clock might suffice.

There the correspondence ends, after lasting eight months, from May to December 1888. It is unclear if any new equipment was purchased, and Christie did not secure any extra permanent staff, at least in the short term. But he did not walk away empty-handed. The Admiralty agreed to increase the budget for boy computers by 40 per cent, allow-ing him to hire eight extra adolescent workers.[132] Of course, without enough supervisors, monitoring them was a challenge. It is likely that this predicament is what led Christie to hire a cohort of educated female computers, including Annie Russell, who were older and more reliable, but could be paid much less than full-time male staff members.[133] In the meantime, Christie continued to badger the Admiralty for more super-visory staff.[134] By 1891, he managed to appoint one new second-class assistant to "strengthen the supervising power" and had secured a prom-ise of more on the way.[135] The promised assistants arrived in 1892.[136]

The time-signal crisis of 1888, though ostensibly a simple labour dispute, brought to the fore questions about the very nature of the Greenwich time service. Christie, like the delegates he had picked to go to the IMC, Adams and Evans, felt that Greenwich time (in its most accurate form, under the guise of the universal day) was a tool for nav-igation and astronomy alone. Watchmakers might use it too, but only because the best of them also made chronometers for the Royal Navy. So when Christie threatened to end the time service, he meant it as a direct attack on the Admiralty. What he did not count on was that the Post Office relied on Greenwich time just as heavily as the Royal Navy. The Post Office made use of the time internally to keep its telegraphy service in order, and secured revenue from selling the time to paying customers. Although the market for accurate universal time was small and special-ized, it was larger than Christie imagined, and growing.

Of course, this still left a hierarchy of times: only the wealthy were able and willing to pay to keep their clocks ticking on time to such an exact degree. Yet there was enough of a private market, beyond just astron-omers and clockmakers, for entrepreneurs to invest in the business of selling time. New companies began to spring up selling Greenwich time.

These firms could not sell it directly. The Post Office had a near-monopoly on access to the time signal direct from Greenwich. Furthermore, those who rented the time from the Post Office had to sign an agreement: "No Electric Time Current or Signal communicated under the terms of this agreement to the said apparatus for the purpose of recording or showing the true Greenwich Mean Time shall be made use of by the said Renter except for his own business or private affairs."[137] In other words, the renter could buy Greenwich time for personal use, but could not sell it without "the written licence or consent either special or general of the Postmaster General."[138]

But private firms became keen to sell the time, and a few made agreements with the Post Office allowing them to do so. One of the first was Barraud and Lunds, clockmakers at 41 Cornhill in the City of London. In the 1870s, it organized a service that synchronized clocks to Greenwich time.[139] Rather than tapping into the telegraphic time signal, it used the Post Office signal to keep a main clock at the company's headquarters on the correct time. This main clock, by way of electric currents, would hourly correct any clocks connected to it by wire. Customers could thus purchase one of these special receiver clocks, and have it connected to the main clock.

In 1882, Barraud and Lunds was broken up, and the clock-synchronizing part of the organization re-formed into the Standard Time and Telephone Company (later simply the Standard Time Company, or STC). This new firm took over Barraud and Lunds's rights to use the Post Office time signal to sell its synchronized clocks.[140] The STC set about developing its own base of customers. It was able to sell the time much cheaper than the Post Office, costing about £4 per year for renters, who also had to pay up front for a special clock. In going about its business, the STC had to be careful not to step on the Post Office's toes. In 1888, the Post Office sought lawyers' opinions as to whether the STC's business model was even legal. The Telegraph Act of 1869 gave the Post Office a monopoly on sending messages via telegraph. However, the lawyers opined in favour of the STC, stating that the clock's electric current did not count as a message. Other, more traditional clockmaking companies offered to tune, wind, and correct their customer's clocks, and the STC was doing essentially the same thing. It did not matter whether the tuning was done by hand or by wire.[141] Indeed, one lawyer suggested that the main clock and receiver clocks could be considered two parts of the same machine, in which case, no "message" of any kind was being sent.[142] The Post Office let the matter drop, and the STC kept selling its synchronized clocks.

It was not the only one. Similar companies popped up through the 1880s and into the 1900s, trying to cash in on the growing demand for accurate, authoritative time. Some of these outfits claimed Greenwich's newfound authority for their own, such as one that unsubtly named itself the Greenwich Time Ltd.[143] Others, instead of hitching their wagons to Greenwich directly, attempted to emphasize their modernity, choosing futuristic-sounding names like the Magneta Company, Remelec, and Synchronome.[144] All appealed to their customers' desire to appear modern and up to date. Magneta, for example, to sell its synchronized clocks, relied on expert testimonials and technical jargon, including a recommendation from William Preece, the Post Office engineer.[145] Others appealed to modernity and progress more directly. As one Synchronome pamphlet put it, "The nineteenth century with all the wonders it has wrought has one scientific disgrace. We still depend for time-keeping upon clocks which require weekly winding and all of which unblushingly tell a different lie."[146]

But Synchronome had the answer. Its electric clocks were supposedly "indispensable in institutions, hotels, banks, offices, &c., factories, and everywhere where loss of minutes means loss of pounds to the employer, and of great value in every household. In schools, a special instrument is added for the control of bells which are automatically rung in the class-rooms at pre-arranged times."[147] In a similar vein, an advertisement for the Greenwich Time Ltd proclaimed that "a cure for unpunctuality, and the host of other tragedies that are supposed to follow the possession of inaccurate timepieces, has been discovered at last."[148] The company compared its time service to other utilities: "It is now possible to have a service of Greenwich time laid on in the home like gas or electric light."[149] This attempt at normalizing the time service as a basic utility was part of firms' broader attempt to expand their customer base: where the Post Office had sold time mainly to clockmakers and amateur astronomers, these companies targeted offices, public buildings, banks, schools, factories, and even the homes of the wealthy.

Magneta received some free advertising in a piece titled "The Romance of a Daily Newspaper" about the *Daily Mirror*, a "lady's penny" paper that had survived low circulation early on to flourish later. The *Mirror*'s offices featured Magneta's synchronized clocks, helping to keep the presses running on time.[150] Magneta clocks serviced several other newspapers, as well as hospitals, the Royal Mint, various postal buildings, and London's fabulous Ritz and Savoy hotels.[151]

Synchronome pitched its services to companies running steamers and ocean liners, as well as a specialized timer for "racecourses, motor tracks, and athletic clubs."[152] Of course, whom it advertised to differed from who purchased these clocks. Actual customer lists included private homes, colleges, breweries (including Guinness), naval barracks, industrial firms, factories, insurance company offices, telephone companies, and city councils.[153]

These companies' expansion of customer bases beyond clockmakers and astronomers is reflected in their ads. One Synchronome ad, for example, targeted wealthy individuals. Taking the form of a picture book modelled on the children's nursery rhyme "The House that Jack Built," the ad (see Figure 4.4) tells the story of "Jack," a house owner who synchronizes his clocks and reaps various benefits, such as a new-found punctuality. A caption under a photo of men rushing for a train leaving a station quips, "This is the train Jack caught in the morn / and left his neighbours all forlorn / cursing the day that they were born / as they thought of their key-wound clocks with scorn / compared with the time that's uniform / all over the house that Jack built."[154] Another line of the poem establishes that the new house clocks regularly "woke Jack's slavey up at dawn," "slavey" being derogatory slang for a female domestic servant.[155] This inclusion suggests the ubiquity of domestic servants among the upper classes of Victorian and Edwardian Britain. More important, it establishes that while only upper-class customers could afford these clocks, working-class Britons, though not paying customers, were nonetheless exposed to the new synchronized Greenwich time in their places of employment. The working classes were users, if not purchasers, of Greenwich time.

Like other ads, "The House That Jack Built" also emphasized Synchronome's modernity, describing its pioneering inventor and its mechanisms, as well as assuaging fears about the dangers of electricity in the home, pointing out that the clock batteries were "hardly sufficient to kill a fly."[156] The final stanzas shake their metaphorical fists at the backwardness of inaccurate clocks (in the accompanying image, "Father Time" shakes his fists at mismatched clocks). "These are the clocks at sixes and sevens / which cost so much and lie – oh heavens! / that have to be wound and are never right / so should not be found in anyone's sight / no wonder that Jack has put them in pawn / and gone in for time that's uniform / all over the house that Jack built."[157] The ad attempted to reach new customers by playing up the anger and confusion over timekeeping that was so evident in contemporary newspapers.

In some cases, new laws inadvertently created customers for Greenwich time. For example, the Licensing Act of 1872 limited the time of day it was legal to sell alcohol, making pub owners into customers of the

These are the CLOCKS
   At sixes and sevens
Which cost so much
   And lie—Oh, heavens!
That have to be wound
   And are never right
So should not be found
   In anyone's sight
No wonder that Jack
   Has put them in pawn
And gone in for the Time
   That's uniform
     All over the House
     That Jack Built

Figure 4.4 | A page from "The House That Jack Built," a Synchronome Company advertisement for synchronized clocks.

time sellers. The Crown Tavern in London purchased the time from the STC in 1884, for a short period at least.[158] Legislation protecting factory workers also spurred industries to adopt Greenwich time. In one case, the Oldham Master Cotton Spinners' Association (near Manchester) requested permission from the Post Office to distribute the time to its members. It wanted, in essence, to "do for their members what the Greenwich Time Ltd. offer in London."[159] It would distribute time, specifically for the cotton-spinning industry, to avoid punitive measures for making employees work overtime. "For many years complaints have been made by our members, in regard to prosecutions by the Factory Inspector for alleged overtime, that the clocks by which the Factory time is regulated vary very much and that in consequence they are on occasion the victims of injustice."[160]

The Post Office rejected the association's request, which would have created another, competing time service, taking business away from the Post Office itself. It suggested instead that individual factories could purchase the time direct from the Post Office. Besides, Greenwich Time Ltd and the STC operated only in London and had existed before the Post Office had begun distributing time. Allowing these two companies to compete with it was one thing, but creating new competitors outside London like the Cotton Spinners' Association was another.[161]

One final customer base that the Post Office and private companies such as Greenwich Time Ltd, the STC, and Synchronome were all vying for business from was the owners of public clocks. Clocks visible from the street were common outside businesses, government buildings, and churches, but there was no guarantee that these clocks were accurate. As a Post Office memo put it in 1913, "There is no doubt a public need for synchronization of clocks – especially those in the streets. It is obvious from the public clocks along Fleet Street and the Strand that some more efficient means of regulating them is desirable."[162] An extract from the *Daily Express* around the same time said: "There will be no defence in near future for anyone who has charge of a public or office clock which fails to keep time," because the Post Office was planning to make its service cheaper.[163] The British Science Guild, in 1908, was also unhappy that there was "no general system by which the public is provided with the means of getting exact standard time ... The Committee are strongly of opinion and think it highly desirable and important that arrangements should be made so that a number of public clocks in different districts of London and other large towns and perhaps the clock at a telegraph office in smaller towns and villages should at certain hours be automatically corrected to agree with the true standard or Greenwich mean time."[164]

Clockmakers, too, complained about faltering public clocks. One clockmaker, worried about his reputation, wrote to the *Horological Journal* that owners of public clocks had a duty to "keep the clock regularly wound and in good repair, so that the clockmaker's reputation should not suffer."[165] Apparently the author had noticed one public rail-station clock that was always wrong and was concerned that the clockmaker would be blamed for its failure, rather than the true culprit, the inattentive owner. Complaints of this nature were widespread in the early 1900s, and some synchronization companies began pushing for legislation to make it illegal for public clocks to show the incorrect time. Such a law would force owners of public clocks to pay one of these firms to ensure their accuracy.

While the synchronizers stood to gain the most from fixing public clocks, many other people complained about them too. But the added pressure from the synchronizers forced the debate into the limelight, in the pages of *The Times* and other major newspapers in early 1908. The STC's secretary wrote to *The Times* in January, suggesting that "the irregularities of London's public clocks are directly responsible for an immense amount of financial loss, in addition to the inconvenience already admitted ... In the present state of affairs every man's time is his own, and no inducement exists for the expenditure of the very small sum which synchronization involves."[166] A flurry of letters followed, some supporting and some rejecting the idea of synchronizing all public clocks by legislative coercion.

The debate culminated in a lecture by STC director St John Winne to the United Wards Club in London on 4 March 1908.[167] Winne commiserated with the writers of the vitriolic letters to *The Times* about unreliable public clocks. It was time to put a stop to the confusion and rid London of all the "lying public clocks." Winne suggested legislation requiring all public clocks to be synchronized. Of course, his company, the STC, was sure to profit, but such a scheme would undoubtedly also be of use to the public and help end the confusion about timekeeping.

Partway through the lecture, Winne made a brief, telling aside about one of his competitors. He was lamenting the backwardness of regular clocks, while – like the synchronization ads – lauding the advances in telegraphy and electrical time signals that made synchronization possible. In his effort to disparage anything unmodern, he provided an example of an old method of accessing Greenwich time, which the STC was meant to replace: "It may be interesting and amusing to some of you to learn how Greenwich meantime was distributed amongst the clock and watch trade in London before the present arrangements came into vogue … A woman possessed of a chronometer obtained permission from the astronomer royal of the time to call at the observatory and have it corrected as often as she pleased. She then made it the business of her life, until she reached a great age, to call upon her customers with the correct time, and on her retirement this useful work was, and even today is, carried on by her successor, still a female, I think."[168] Some members of Winne's audience, which included several clockmakers, were aware of these women. Daniel Buckney, who worked for prominent clockmaker Dent and Co., confirmed Winne's story.

On the whole, Buckney was unhappy with Winne, as were most other clockmakers in the room. The suggestion that public clocks needed synchronizing carried with it the implication that clockmakers were inept and that their clocks could not reliably keep the time. So the clockmakers were keen to rebuke Winne and his synchronization scheme. Their responses to his lecture were largely negative, but they too were dismissive of the women he spoke of, because these women were also in the business of correcting the errors of their clocks. Buckney told the audience, "It is quite true, a lady did do it [delivered Greenwich time], and another took her place, but I may say that that lady calls at our establishment to see whether she has the correct time (laughter)."[169] Buckney then insulted Winne's company: "The synchronizing company receive the signal from Greenwich by our standard clock. (renewed laughter)."[170] There was clearly competition between the people making the clocks and those synchronizing them. But equally significant for our story is the way Winne talked about his competitors, these unnamed women.

Winne made them objects of ridicule. He labelled their time service "unofficial," with the connotation of illegality, or at least of unreliability. And he described their business in the past tense, before electricity. He was presenting them as anti-modern things of the past, and his own electric clocks as the way of the future. According to one newspaper report, Winne also suggested that the women used their feminine wiles to gain entry to the observatory each week, which "perhaps no mere man could have got."[171]

Who were these women whom Winne was so eager to discredit? The newspapers covering his lecture in 1908 were eager to find out and hunted them down for an interview. They discovered Ruth Belville, who was still operating the business she had taken over when her mother, Maria, died in 1899.[172] There is a substantial archival record of both of them. Maria's husband, John Henry, had worked at the Greenwich Observatory in the 1830s. He was tasked with setting up a service by which a chronometer was corrected to the right time at the observatory,

Figure 4.5 | Maria Belville.

and he then carried it around town to watchmakers and businesses keen to learn the accurate time. Maria Belville took over this task after his death in 1856.[173] As a single mother with a young daughter, Maria sometimes took Ruth along on her route. In fact, one of her husband's wealthy friends offered to give the girl an education, but Maria declined, saying that she had her own small income and didn't want her daughter taken away. So the two stayed together and continued the time service that Ruth's father had begun.[174]

The number of customers the Belville family served varied considerably in the century, between the 1830s and the 1930s, in which they sold the time. Ruth estimated that her father had over two hundred, while she herself maintained around fifty late in her career.[175] Most were watchmakers, but they also included factories, shops in fashionable parts of London, and millionaires' homes.[176] These figures also omit a sizable number of secondary customers. Ruth wrote later in her life that she remembered visiting a large clock firm in Clerkenwell with her mother. When the two women were leaving after their delivery, they passed three or four people coming in, watches in hand. These people, Maria explained, could not afford the Belvilles' fee, so paid that firm a smaller one to take their time second hand.[177] So the Belvilles supported a network of Greenwich time users much larger than their immediate customer base.

The Belvilles charged about £4 a year for a subscription, just slightly more than the STC, but much less than the Post Office.[178] Maria retired in 1892 (see Figure 4.5), and Ruth replaced her, selling the time into the late 1930s. Their once-weekly route (on Mondays) to their customers varied over the decades. They covered part of it on foot, but they relied on multiple methods of transportation. In the early years, Maria did her rounds on boat taxies on the Thames. Later, when more of London's rail infrastructure was in place, she rode on trains, trams, and buses.[179] Even still, they would have been long days, especially for Ruth, who later in her career moved out of London to a smaller town, Maidenhead, near Windsor, and had to commute – about 38 miles to Greenwich and 27 to the City of London. Still, the business made up much, if not most, of their income.

Both Maria and Ruth's census records say nothing of their distributing time.[180] Maria listed her occupation as schoolmistress, and Ruth governess. This does not necessarily mean that they regarded their time business as secondary to their primary professions, but suggests that governess or schoolmistress was a more respectable position to enter in a government census than "purveyor of time." They maybe feared

Figure 4.6 | Ruth Belville poses in front of the Royal Observatory, Greenwich.

losing their special access to Greenwich time if they announced themselves to the wrong government body. After John Henry's death in 1856,
they were not in any way employees of the observatory. The astronomer
royal knew of their weekly visits to the Greenwich Observatory and
their time-distribution business, but it is possible that his superiors, the
Admiralty Board, did not. The closest Ruth went to revealing her business in the census was in 1901, when she listed her profession as "living
on own means," but did not provide any specific details.[181]

Their reticence is not surprising. Both women relied for their business
entirely on the goodwill of the astronomer royal at Greenwich. When
Maria's husband died in 1856, she applied to George Airy for a pension
from the Admiralty as a widow of an employee.[182] The Admiralty denied
the request, as the spouses of civil servants were not entitled to a pension.[183] Maria persisted, however, first asking if the observatory might
buy her husband's scientific papers and collection of weather journals.[184]
This request was denied too, although she did eventually find a buyer
for them. Airy's letters make it clear that he wanted to help Maria, but
that these decisions were out of his hands.[185] Her last hope was that he
would allow her to take over her husband's time business. She wrote:
"I am encouraged by your goodness to advance another petition. Being
engaged to take the Greenwich time to 67 of the principal chronometer makers in London I have to request admission once a week to the
clocks in the observatory in order to test my own regulator – it would
inspire those who have taken up the widow of their esteemed friend with
additional confidence if you could accord me this favour."[186] This was
something Airy could agree to on his own without having to consult the
Admiralty. And so Maria was allowed in.

But her position was insecure. A few weeks later, she was accused of
breaking into the observatory unannounced and leaving the gate unlocked.
Airy assumed she had a key that her husband had not returned and had
used it. He demanded she return the key and enter the grounds only via
the main gate by asking the porter.[187] Maria wrote back apologetically,
saying she had no key, and that she had found the gate open. Someone else
must have forgotten to shut it.[188] Airy confirmed with his staff that John
Henry's key had been returned, and he let Maria off the hook.[189]

Thirty-six years later, in 1892, when Maria retired, Ruth wrote cau-
tiously to the new astronomer royal, William Christie, for permission
to carry on her mother's business.[190] He agreed, and she continued
selling the time around London.[191] Again, however, there was trouble.
Maria's retirement caught the attention of the newspapers. A *Daily
Graphic* article sought to explain why the Belville service existed at all,

as it seemed redundant. But the journalist added: "It is a well known fact that" the Post Office's time signal via telegraph wire "is not to be relied on."[192] Senior staff members at the Greenwich Observatory were not happy with the insinuation that the time signal they sent to the Post Office was inaccurate.

The observatory's chief assistant wrote to the paper to complain, saying that Post Office time was perfectly accurate. The Belvilles' time service, he indicated, was merely an artefact of an older age, before telegraphic arrangements had been made. "Her present usefulness, I believe, is in supplying the approximate time to those who find the Post Office charges too high," he wrote, suggesting that her clock was only "approximate."[193] Instead of complaining, however, Maria wrote quickly to the observatory and apologized profusely, making it clear that the paper's misstatements about the quality of the Post Office's time signal did not come from either her or her daughter.

Because of the precariousness of her access to the observatory, Ruth Belville must have been incredibly nervous in 1908 when she again became the focus of media attention (see Figure 4.6). After STC director St John Winne's lecture to the United Wards Club, newspapers tried to hunt her down for an interview. The *Daily Express*'s first headline, like Winne, talked about Ruth's time service as a thing of the past: "Woman Who Sold the Time," the headline read.[194] But then the reporter found Ruth and requested an interview and a photo. The second headline moved into the present: "Woman Who Sells the Time: Strange Profession of the Belville Family."[195] So although the headline changed to the present tense, it now talked about the Belvilles' job as a "strange" one, something out of place, an anachronism in the modern world. Other newspapers followed this lead. The *Maidenhead Advertiser* labelled her job "a unique position," while the *Kentish Mercury* wrote: "Greenwich Clock Lady: Romance of a Regular Visitor to the Observatory."[196] Here, again, "romance" invokes nostalgia, as if Belville was out of place and time.

Let us consider the language used. Why did the papers consider selling the time a "strange trade" or a "strange profession"? Coverage of Winne's lecture, the STC's synchronizing business, and the Post Office's time signal did not use the same language of "strangeness." No one accused Winne of being an anachronism. And yet the newspapers considered selling the time a strange profession when Maria and Ruth Belville did it.

There are two explanations for the tone. The first relates to gender. Belleville's profession fell outside accepted gender roles. This perception of abnormality prompted attacks against her morality, such as a suggestion that she seduced the astronomer royal to gain access to the observatory.

Belville's respectability was questioned much in the same way as suffrag-ettes' morality was being questioned in the same period. Timekeeping in the era of Greenwich mean time was considered a science, which was largely a masculine pursuit. The Belvilles were not welcome in the new "profession" inhabited by the electrical-clock experts at the STC.

The second explanation for the "strangeness" of the Belville time ser-vice concerns Edwardian notions of modernity. Selling the time using electricity was not strange, but doing so by hand was. In the Edwardian vision of progress, new technologies did not just supplement old meth-ods, they replaced them entirely. There was no room for the Belvilles alongside the STC. The contrasting treatments of these time-sellers reflect a patriarchal and unchallenged belief in progress, the separation of spheres, and the pre-eminence of a vision of modernity that valued technological solutions to everyday problems. Winne's rhetoric, echoed in the newspapers, capitalized on these values to undermine the Belvilles. It turned them into a spectacle rather than a serious competitor in the time-synchronization business. Winne and the papers focused on both gender and imagined temporal spaces to delegitimize the Belvilles, push-ing them back into an imagined past. The time-saleswomen were sup-posedly antiquated oddities, inaccurate and unscientific. In contrast, Winne and the STC with their electric wires were apparently masculine, modern, scientific, and perfectible.

The stigma was hard to shake. Similar headlines attached themselves to Belville again in 1913. The *Daily News and Leader* trumpeted, "Lady Who Has Inherited a Strange Trade."[197] Also in 1913, the *Observer* again relegated Belville's time service to the past, printing a story about "The Belville Tradition."[198]

But Ruth did resist. In a few places, we catch rare glimpses beyond the modernist filter of the STC and the newspapers, and Ruth speaks in her own voice. And it is quite the contrast. Ruth rejected the narrative of Winne and the papers that relegate her to the past and suggest that her time service was unreliable. She told her local newspaper, the *Maidenhead Advertiser*, that, despite Winne's claims, the biggest clockmaking firms "will have nothing to do with synchronized time or any other means of communicating the time," besides herself.[199] To these firms, her method was tried and true. She had earned their loyalty. Ruth also publicly replied to the newspaper that had called her profession a "tradition":

> Sir, my attention has been called to a paragraph in the article headed "The Correct Time" in the number of *The Observer* for the 24th instant, which is headed "The Belleville Tradition." I take objection

to the word "tradition." Tradition means a thing of the past, that has been handed down orally from generation to generation. My distribution of Greenwich Mean Time takes place at present. The error of my chronometer is ... stated on a written certificate; not written and signed by me but by unimpeachable authority. [The] weekly error of my chronometer ... seldom exceeds a few tenths of a second. As to synchronized clocks, doubtless they are of service to the general public and possibly to those who sell cheap watches, etc, but to the high class scientific watch and chronometer maker, Greenwich mean time is required to tenths of seconds; and this can be provided by no better means than that of a first-class English chronometer, accurately adjusted and timed to tenths of seconds from the Royal Observatory at Greenwich.[200]

Ruth was aware of the game Winne and the papers were playing, and she firmly rejected it. She later wrote privately: "I think the Standard Time Company will not attack me again in public as the result ended in rather a heated discussion at the end of the lecture and the last thing that Mr. St. John Winne wanted was to advertise the chronometer [which she carried] at the company's expense."[201]

Ruth was probably right about the quality of her service matching, even exceeding, Winne's. Her customers could rely on her time service without worrying about telegraph-line failures, delays, and interruptions that plagued the STC and the Post Office.[202] The notion that the Belvilles' time service remained relevant in the electrical age was first articulated by historians David Rooney and James Nye. They argue that the Belvilles' service, far from being backwards and outdated, was considered much more reliable and trustworthy than the supposedly modern STC.[203] "New technology," Rooney writes, "doesn't just sweep aside old systems. They coexist far longer than one might expect ... From the users' perspectives [the STC and Post Office's time service] was good but not always good; available but not always readily so; accurate enough for most people most of the time but no more so than Ruth's service."[204] The records of the Post Office and the Greenwich Observatory corroborate this claim. The wire to the time ball at Deal (on the Kent coast, where the Thames meets the Channel) was occasionally faulty, as was the wire to Westminster and Big Ben.[205] The astronomer royal had to field as many complaints as the Post Office did about time-signal failures, most often because of problems with the wires.[206] This problem was not unique to London. Charles Piazzi Smyth received many complains about his public signal, the time gun in Edinburgh.[207] The British observatories in Cape

Town and in Durban (Natal) had similar technical failures.[208] In London, one engineer inspecting the Greenwich time-signal wires in 1887 went so far as to say that, because of the poor state of the public time-signal apparatus, "it is surprising that failures are not more frequent."[209]

The new technology for synchronizing clocks was expensive, and it often failed. Indeed, by the time the technology was wholly reliable, it was becoming outmoded, giving way to the more accessible and cheaper wireless-radio signals. A Post Office memo from 1915 proclaimed that "the need for extreme accuracy in time keeping is, comparatively, so small that it is scarcely worthwhile for the department to take up the matter seriously or to expend any large sum of money for the purpose of safeguarding its monopoly. Moreover, the demand for this particular system of time signalling has been diminished by the distribution of standard time signals by wireless from the Eiffel Tower."[210]

Despite whiggish notions of progress, technological advances are never straightforward. Ruth and Maria Belville's hand-carried pocket watch was not replaced by the STC's wires, or even by the radio, at first. Rather they existed in parallel, with Ruth Belville's time service continuing well into the 1930s. Of course, it could exist only in a place like London, with a high concentration of customers and direct access to the observatory. The Post Office's wires, in contrast, opened up a broader system of time regulation, reaching across all of Britain's cities. Rural timekeeping was more fluid, but the network of wires and rails slowly reshaped the temporal landscape of the country.

CONCLUSIONS: THE CLOCKS ARE TELLING LIES

Attempts to standardize time globally did not simplify timekeeping. Instead, they created new layers of complexity, confusing the public about whose time was correct. Access to accurate universal time was hierarchical and unequal. Accurate time was expensive, and urban, and therefore limited to professionals and those who could afford it as a luxury good. These problems came to a head in the case of the STC's and the Belville's time services, raising questions about whose time was authoritative, and whose was not. If we look at the language used by letter-writers to *The Times*, clocks not set to Greenwich were not simply showing a different time, they were "lying." Yet only the wealthy could afford time that told the so-called truth. In a similar vein, attempts to delegitimize the Belvilles used language that linked authority to both modernity and masculinity, thus relegating their time service to the imagined past, an object of nostalgia and romance. To assert their new temporal authority, telegraphic time companies like the STC had to delegitimize their competitors. Similarly, if Greenwich time was to reign supreme, local time had to die. Its emergence as the "true time" undermined the authority of local timekeeping methods and removed agency from peripheral timekeepers.

The Belvilles occupied, if we may quote from one of their newspaper headlines, a "unique position" vis-à-vis the constructed authority of Greenwich time. On one hand, they were part of the machinery that legitimized Greenwich's authority: they were keen to show that the time they proffered was indeed the best, the "true" time, to the detriment of more affordable versions of public time. On the other hand, just as anarchist Martial Bourdin's attack on the Greenwich Observatory was a rejection of Greenwich time, Ruth Belville's weekly rounds on foot, carrying a chronometer, were in some ways just as powerful a form of resistance against the constructed authority of patriarchal modernity and uncritical technological progress.

This chapter has made the case that the inaccessibility of scientific time meant that most Britons did not begin using it for decades after it first became available, even mocking it to the point of ridicule. Yet at the same time, it could be appealing, not to mention lucrative, to adopt the trappings of scientific time. It allowed users to appear modern, legitimate, and forward-thinking. This central tension drove most of the conflict in this period over how to deliver scientific time and to whom. The IMC's "universal day" divided, rather than united people, because of its uneven implementation. The new timekeeping system served to reinforce social inequalities. As we saw with the Belvilles, however, and as we see again in the next chapter, there was room for resistance and room to reclaim authority over time by those outside the scientific community.

5

Teaching Time, Using Time

In North America as in Britain, the public's relationship to universal time was complex. The International Meridian Conference (IMC) of 1884 had decreed that the universal day would not affect civil time, but Canadian and U.S. railways already ran on standard time, which used the Greenwich meridian and therefore derived from universal time. Yet the same technological limitations and high cost of distribution of accurate time that plagued Britain also affected North America, resulting in uneven access to accurate standard time beyond the railways. "Time-sellers" and their wealthy customers championed accurate timekeeping as a symbol of their modernity, but they were few. Most people, unable to pay for the luxury of scientific time, ridiculed its superfluous precision.

All of this was true on both sides of the Atlantic, yet the two worlds experienced time distribution differently. While Britain maintained a single timekeeping authority at Greenwich, the United States experienced a much more decentralized process of time modernization, under the auspices of various privately funded universities and observatories across the mid-continent. Canadian timekeeping was similarly managed by various regional authorities, although these were government-operated institutions rather than private ones, and they were slowly collected into a single body by the turn of the century to more closely match the British model. But the key difference in timekeeping practices between Britain and North America was the scale. North America's longitudinal width had necessitated the use of multiple time zones, leading to more diverse timekeeping practices. Thus efforts to unify and standardize timekeeping required not just better access to standard time, but also a massive education campaign to reprogram the public's timekeeping behaviour. Reluctant people needed to be brought around to standard time, if not by legal coercion, then by re-education.

Time reformers like Sandford Fleming sought to do just that. Time reform and time distribution in North America became closely tied to education in the late nineteenth century.

This chapter uses the lens of education to focus on questions of power and authority – about who had the "right" to decide the time. We explore this issue through several case studies. The first is a conflict between U.S. private universities and government institutions over who had the authority to distribute accurate time. The second shows how efforts to instil new timekeeping practices via government coercion/law failed in Canada, so reformers turned instead to education to establish new timekeeping norms. Finally, the chapter ends with a case study of how some Indigenous peoples in Ontario, such as the Council of the Delaware Nation at Moraviantown, used standard time to assert their own political authority in the face of assimilationist policies. In all these examples, it is clear that controlling the source of authoritative time-keeping was a valuable political tool, and could be applied to assert dominance – or undermine it.

Education itself was undergoing significant reform in this period. As it became more widely available to more people, debates raged over its scope and purpose – which subjects ought to be taught to whom, and by which methods? In Canada, the reform debates centred largely around languages of instruction (French/English) and the place of religious teaching (Catholic/Protestant) in public schools. Reformers such as Egerton Ryerson (the long-time superintendent of schools for Upper Canada/Ontario) supported secular education in order to prevent one denomination overshadowing others in public-school curricula.[1] Further debates centred around whether schools should teach academic subjects, or more practical industrial and agricultural skills. The turn-of-the-century Macdonald Robertson Movement, for example, spearheaded by philanthropist reformers Sir William Macdonald and James Wilson Robertson, attempted to consolidate small rural schools into larger institutions, at which agricultural skills rather than "esoteric" arts would be the focus of the curriculum. This period also saw education become available to more people than ever before, with school becoming free and compulsory in most provinces by 1910, beginning with Ontario in 1871 and finally Quebec in 1943.[2]

Educational reform in this period was often based on middle-class assumptions concerning social ills and the rehabilitation of the poor. School reformers saw education as a way to improve and "civilize" human societies, pulling "street urchins" into classrooms, preventing crime, and giving impoverished youth a productive future.[3] Yet reformers were often unaware that their curricula were not neutral, nor evenly available. Class, race, and gender shaped the quality of education available to children. In Canada, schools were also colonial institutions. As historian David Willinsky suggests, imperial education "gave rise to peculiar and powerful ideas of race, culture, and nation that were, in effect, conceptual instruments that the West used both to divide up and to educate the world."[4] Curriculum reinforced racial hierarchies. The world: its geography, its history, its land and resources, humanity itself, were all classified, organized, and slotted into an imperial worldview. Schools entrenched imperial hierarchies in students' minds.

The way timekeeping was taught in schools was no less problematic. A particularly western, Greenwich-centred time-sense was inculcated through curriculum and through the very structure of the day. Schools were the instruments by which the norms of public timekeeping were cultivated, shaped, and challenged. At higher levels of education, especially in the United States, educational institutions were the literal producers and distributors of accurate time for their communities, via privately funded university observatories that sold time to railway lines. In elementary and high schools, meanwhile, children were taught new time systems via formal curricula in hopes that they would in turn teach their parents, raising a generation of "modern" timekeepers. Meanwhile, schools instilled a sense of clock time through their schedules and bells. Reformers hoped that these programs might create a new, modern time-sense in children. But these ambitions were only partially fulfilled. Students' understandings of timekeeping differed by context. Students in urban centres adapted quickly to standard time, while many rural populations remained indifferent. Indigenous children and their parents, meanwhile, experienced time reform in the colonial environment of both residential and day schools, but also found ways to adopt and repurpose it to their own uses.

Regardless of how students' experiences varied across the continent, education was universally one of the most pervasive processes through which civil society engaged with timekeeping practices. As a place where the findings of science, combined with the state's ideological beliefs, filtered into public consciousness, schools are ripe subjects for examining the disconnect between the scientific production of knowledge and civil society's practical use (or rejection) of that knowledge. Timekeeping is no exception, and the way it was used and taught in education offers a glimpse at how controlling timekeeping offered an indirect means of wielding power and authority.

THE UNIVERSITY AS TIMEKEEPER IN THE UNITED STATES

Time measurement in the 1880s United States could be frustratingly confusing, more muddled than in Britain. Britain had the luxury of "owning'
the prime meridian. Its universal time was Greenwich time, which was
railway time, which was legal time. These categories did not line up so
neatly in the United States. Instead, as we've seen, the railways adopted
a series of one-hour time zones in 1883 to account for the continent's
unwieldy east-west expanse. But this was, at first, mainly a specialized
time for travellers. It was not clear whether the public was meant to
adopt railway time in everyday life, and many people did not. Whole
cities continued with local time, as did most rural areas.[5] The federal
government, meanwhile, maintained a laissez-faire policy towards the
subject. The IMC did little to clear up the matter, failing to enshrine standard time in international law, although it did enshrine Greenwich as the
system's prime meridian. The challenges of accessing accurate standard
time exacerbated this confusion. Distribution was expensive and consequently limited, and so created a hierarchy of times, in which standard
time was a luxury for the wealthy, a specialized tool for the professional,
and a nuisance to railway travellers, but was not otherwise widely available. Most people, should they desire it, were dependent on the uneven
chance of living near an accurate public time signal or train station or
working for an employer who wished to instil punctuality.

With this multiplicity of times across a vast geography, constructing
a single, unrivalled authority on time was difficult in the United States.
Britain's highly centralized time-telling infrastructure based on the
observatory at Greenwich had no equivalent in the United States. The
superintendent of the U.S. Naval Observatory in Washington, DC, eventually gained some semblance of authority over the nation's timekeeping, but never equalled Airy or Christie's status at Greenwich. Finnish
astronomer Anna Molander, writing to an employee of the Greenwich
Observatory in 1909, explained that U.S. astronomy was very different
from Britain because "there are in this country [the United States] so
many private observatories."[6] Dozens of these private observatories
measured time individually for their corners of the country, each independent of the others. Nearly all of them were operated by colleges and
universities, which established direct links between higher education
and timekeeping.

University observatories had begun distributing the time to paying
customers as early as the 1860s. Some – Harvard and Yale – obtained
a direct connection to Greenwich time after the second transatlantic

telegraph line was completed in 1866 (the first had failed months after installation). But most universities had to produce/measure standard time themselves via transit astronomy. These sources introduced a level of error that varied by place and over time, so American timekeeping was diffuse and operated in patchwork.[7] The U.S. Naval Observatory operated a time ball in Washington and sent the time to New York by wire. But elsewhere, academic institutions kept the time.[8] Some of the most active, besides Harvard and Yale, were the University of Cincinnati and the Alleghany Observatory at the University of Pittsburgh.[9] Each developed its own time-distribution networks, which functioned alongside companies – analogous to Britain's Standard Time Company (STC), the Greenwich Time Ltd, and Synchronome – that began to emerge in major cities, such as New York's Self-Winding Clock Co., which distributed U.S. Naval Observatory time to customers for $12 a year.[10]

Unlike in Britain, timekeeping was a profitable enterprise for U.S. observatories, not just for the distribution companies. Being a producer of knowledge – in this case, a producer of accurate, authoritative time – was an opportunity for making money. In Britain, Greenwich had a near monopoly, but the Post Office reaped any income earned from selling the time. In contrast, U.S. university observatories competed with each other for customers. Selling the time subsidized research, as revenue from time distribution supported researchers' scientific endeavours. As a result, academic astronomers had more of a stake in cultivating and preserving a customer base. While Astronomer Royal Christie had felt perfectly comfortable threatening to end the Greenwich time service altogether, calling it "extraneous work," in American universities that same type of time-distribution service was considered essential to the observatories' continued well-being.

The reliance of university observatories on their time services for research funding incited a protracted conflict in 1883 between the Naval Observatory and the observatory of Washington University in St Louis, Missouri. The catalyst for this conflict was the 1883 decision by American and Canadian railways to adopt time zones. In order to ensure a smooth transition to the new system, William Allen had written to Robert Shufeldt, the superintendent of the Naval Observatory, on 6 October 1883, informing him of the plan and asking for his cooperation.[11] He was worried Shufeldt might refuse, as, in the past, the observatory staff had preferred to use the meridian of Washington, not Greenwich.

Allen need not have worried. The plan would simplify time distribution for the Naval Observatory. Although Shufeldt would have preferred a single universal time instead of twenty-four different time zones, he considered the system a "move in the right direction."[12] He saw the new railway time zones as a temporary solution, to be inevitably "subjected to criticism and a better plan evolved from it."[13] But in the meantime, he would support it. Shufeldt not only agreed to distribute zone time for the railways, as requested, but also promised that "unless there be some unexpected opposition, [I will] try to secure the immediate adoption of the same time as the local time for the whole section in which it is to be used by the railroads."[14] The observatory, in doing so, would relieve itself of the task of calculating the numerous local times that it currently provided. As Shufeldt explained, "By the proposed plan of having standards differing by one hour, it is made possible to furnish the mean time of each standard meridian by a single time signal; for the signal which marks noon of the 75th meridian, would mark the hour of eleven for the meridian of 90 degrees, and so forth."[15] One time signal for all meant far less work for the astronomers.

In fact, the Naval Observatory officials were so enthusiastic about the plan that they announced that they would "furnish the time, free of charge, to any other telegraph company [besides Western Union, to which they already supplied the time] that will bring a wire to the Observatory."[16] A representative repeated this offer at a meeting of railway managers in Chicago later in October. From this meeting, academic astronomers got wind of the plan, and immediately opposed it.

The most outspoken critic was a professor of astronomy at Washington University in St Louis, Henry Pritchitt. Pritchitt had been part of the international project to document the transit of Venus, travelling to New Zealand to observe the phenomenon in 1882. Such expeditions were expensive, and Pritchitt relied on the profits from his observatory's time service to fund his work. Upon his return to St Louis, he found that income in jeopardy. Railway-manager customers who had previously purchased the time from his observatory in St Louis were threatening to take the time from the Naval Observatory instead, unless Pritchitt provided it free of charge.[17]

Pritchitt's knowledge of what had been said at the railway meeting in Chicago was second hand. As he understood it, the Naval Observatory had announced that "the time signals would be sent free over the entire United States from the Naval Observatory, that the sending of the signals from the observatory was an essential part of the standard time scheme and that for certain reasons the Naval Observatory was the one above all others which should do this work."[18] The facility's representative, according to Pritchitt, had given the railway managers "impressions very unjust to the private observatories of this country."[19]

"Such statements as these," he continued, "are unworthy [of] the national Observatory and unjust to those private observatories which have at great expense of money and labor built up time services in various parts of the country."[20] Why, Pritchitt demanded, was the government-funded Naval Observatory setting itself up to compete with private observatories, which relied on the income for their scientific work? The Alleghany Observatory (University of Pittsburgh), the observatory at Harvard University, the Morrison Observatory (Pritchett College, Glasgow), and the Washburn Observatory (University of Wisconsin-Madison), among others, all stood to lose the same benefits.

The superintendent's reply to Pritchitt was defensive. He did not deny the statement, but tried to underplay its significance: "The Naval Observatory does not propose to make any changes in its time service beyond such as are necessary to conform to the new standards of time, should they be adopted ... The Naval Observatory has distributed time daily by telegraph for twenty years ... No charge has ever been made for this work by the Observatory."[21] It had always provided the time to Western Union for free, argued the superintendent. Any railways wishing access to the time made arrangements with Western Union, rather than with the Naval Observatory directly. Shufeldt argued that Pritchitt had already been competing with the Naval Observatory for years. The only change was that the adoption of standard time had been good advertising for the Naval Observatory, costing Pritchitt his customers.

The Naval Observatory's intentions may not have been insidious, but its actions had consequences. Pritchitt wrote again a few weeks later: "I fear that a considerable part of our income has been permanently cut off and it certainly has been cut off for the present."[22] Pritchitt could not deny the Naval Observatory's right to continue giving the time to the telegraph company for free, but he did take issue with the other insinuations made by its representative at Chicago: first, that the Naval Observatory was somehow "better fitted" to deliver the time than private university observatories; second, that it could send the time anywhere in the United States accurately and regularly; and third, that in the new standard-time scheme only one observatory could send the time. Pritchitt asked whether these statements, which he had heard at second hand, represented the Naval Observatory's official position. If not, then Pritchett could correct his recently lost customers on these matters.[23]

The response was unrepentant. "I considered my reply to your first letter as a complete answer to your questions," Shufeldt wrote.[24] The Naval Observatory had been asked to provide the time signal in cooperation with the standard-time scheme, and it would do so to anyone who made the proper arrangements with a telegraph company. The letter concluded, "While I regret that the distribution of time from the Observatory may diminish your revenue derived from such work, I cannot on that account withhold the cooperation of the Observatory in a matter of so much public importance."[25] Pritchett and the other educational institutions were out of luck.

Of course, the Naval Observatory did not immediately become the centre of U.S. timekeeping in the way that Greenwich did in Britain. Universities continued to sell the time to customers even after the switch to standard time, and U.S. timekeeping remained highly decentralized. But over the next few decades private observatories found selling their own time signals increasingly unprofitable. In 1888 there were at least twelve private observatories selling the time, but four year later, in 1892, that number was down to eight.[26] Pritchitt struggled on, taking up his grudge against Western Union and the Naval Observatory again in the early 1890s, but there was no going back.[27] Most private observatories wound down their time services by 1900. Only a few lasted until the First World War and beyond. But by then, the Naval Observatory's prominence was clear.[28]

In the United States, accurate standard time as a scientific tool, as well as a source of academic funding, eventually lost out to the notion that accurate time was a public good. But that transition came late, and met with significant resistance. In an interesting reversal from Britain, government institutions in the United States were the instrument of establishing free timekeeping (although access was still far from universal), rather than enforcers of a paywall like Britain's Post Office. Instead, it was U.S. academic institutions that claimed, and fought hard to keep, the right to educate the public on proper timekeeping practices, for a price. Private universities, in the business of selling knowledge to the affluent classes and business interests, intended to act as the gatekeepers of temporal knowledge, but were eventually undermined by the drive to make the time free and universal.

TIME DISTRIBUTION IN CANADA

Canadian timekeeping practices mirrored British and U.S. developments. The same confusion existed as to which time was to be used and by whom and how the reforms of 1883 and 1884 were meant to be implemented. As for the distribution of accurate time, Canada resembled the United States more than Britain, at least at first. Where Britain had a single authoritative producer in the Greenwich Observatory and its astronomer royal, geographical realities decentralized U.S. and Canadian accurate time. It was only after the turn of the century that Canada began to consolidate timekeeping under the Dominion Observatory in Ottawa, established in 1905. Until then, some producers of time claimed to be more correct than others, especially when railways standardized time in 1883, threatening the authority of local time.

Educational institutions played a smaller role in time distribution in Canada than in the United States. Government-run observatories, not universities, were the primary sources of time for Canadian cities, and they had a difficult task. Distributing accurate standard time across the whole country required careful regional coordination. As a relatively new conglomeration of smaller colonies, Canada was home to quite a few independent time signals. Halifax, Saint John, Fredericton, Quebec, Montreal, Kingston, Toronto, Vancouver, and Victoria all boasted their own timekeeping observatories at some point after 1850, although they sometimes coordinated with each other.[29] The Halifax Citadel's noon gun and time ball, for example, were operated via a telegraphic link from the observatory in Saint John.[30] In Montreal, observatory director Charles Smallwood operated a daily time ball at the city's shipping wharf. The Montreal Observatory also gave "time to the city by means of the fire alarm telegraph wires" and transmitted local time to the Post Office in Ottawa.[31] The Toronto Observatory similarly operated time signals for Toronto, while overseeing the time service of the observatories in Montreal, Quebec, and Saint John.[32]

When, in 1905, the new Dominion Observatory in Ottawa finally began providing Canada's primary time signal, it subsumed the other observatories under its direction. On top of connecting the time services by wire across the country, it also set up a sophisticated time-distribution service for the Ottawa area, for government buildings in particular. Synchronized clocks were installed in the Parliament Buildings, and by 1907 the system included 227 clock dials, with plans to extend it to the Post Office and the Mint.[33] The Dominion Observatory operated a time gun and time ball in Ottawa as well.

Like elsewhere, the cost of these time signals, and who should bear it, were matters of some debate. Unlike the private U.S. observatories, Canadian observatories charged no fee for their time services. In 1909, this fact irked some of the astronomers at the Toronto Observatory, whose time service was quite extensive. They wrote to the astronomer royal at Greenwich, asking "what charge, if any, you make to the City of London or any private corporations for giving the time. Our Time Service in Toronto is assuming very large proportions and heretofore we have made no charge, a service which today is being somewhat abused."[34] If the Toronto astronomers were hoping to find a precedent for charging for the time service, they were disappointed by the astronomer royal's reply, which pointed out that any income from the time service in Britain went to the Post Office.[35]

The observatory in Quebec City faced similar challenges. Quebec was a major port, so time signals for the harbour were vital to the city's economy. The observatory had implemented a time signal as early as 1856, with a small transit room for observing star transits and a ball tower at the Citadel. But Edward David Ashe, the superintendent, wanted to expand the observatory's function to include longitude determination, meteorology, and the discovery of celestial phenomena, in essence turning Quebec's simple time observatory into a world-class, multi-purpose astronomical observatory.[36] The Canadian Institute, still young at that time, supported his request.[37]

Under Ashe's direction, the Quebec Observatory did expand, but timekeeping remained central to its purpose, especially for mariners. As in any large port, in Quebec shipping interests made frequent use of the time signal. Ashe wrote to George Airy in 1869, asking about electrifying the time ball, and received detailed instructions on how to do it, particularly with reference to a cold climate. (In winter, the Quebec Observatory did not operate its time ball, instead firing a time gun. Each spring, when the ice cleared and the St Lawrence River was navigable again, the time ball was put back into operation.)[38]

Edward David Ashe's son, William, took over the observatory in the mid-1880s, continuing his father's focus on meteorology and exploratory astronomy. But the younger Ashe occasionally neglected the time signal, and he was scolded in 1888 for taking his transit measurements too often by the sun (less accurate), not correcting the error of his instruments carefully enough, and not taking star transits often enough.[39] As in Britain, where time signals were often unreliable, complaints about inaccuracies in the Quebec Observatory's time signal were common.[40]

Several incidents surrounding William Ashe's time service raised serious questions about time signals in Canada. Should they be a public good easily accessible to everyone, or a commodity for professionals and paying customers only? This debate played out in a dispute over whether the government ought to pay for advertisements promoting the time service. In 1889–90, the dominion Department of Marine and Fisheries began to investigate how much other countries spent on such ads. It wrote to its British and American counterparts, asking if their governments paid to advertise their time signals to the public.[41] The U.S. Naval Observatory replied that newspapers reported their time signal free of charge, since it was of great interest to many readers.[42] The British Admiralty's response implied that its time signals were not advertised much, although mariners could purchase a pamphlet describing them at major ports worldwide.[43] The Canadian department understood from these replies that "it is not customary to pay for the advertising of time ball notices."[44]

The department also learned that in the United States, Western Union distributed time for free from the Naval Observatory. Armed with this information, it reached out to Canadian Pacific Railway Telegraph, which agreed to do the same, from domestic sources, in Canada.[45] A few weeks later, however, the company withdrew its offer, after learning that the time signal would be sent at noon, which was peak time for telegraph traffic.[46] The Great North Western Telegraph Company also declined, saying that the great distances involved would make it "difficult, if not altogether impossible, to perform the Service at all to your satisfaction … I am sure it would only lead to disappointment and dissatisfaction to all concerned."[47] Just as with the STC's wired clocks in Britain, time delivered by modern technology was not always the most reliable.

Nonetheless, the producers of time had to find a way to establish public confidence in their ability to deliver it. In 1891, after a shipmaster complained in Quebec's *Morning Chronicle* about the lack of timekeeping data available to mariners, William Ashe wrote to Charles Carpmael, director of the Toronto Magnetic Observatory and superintendent of the Dominion Metrological Service at the Department of Marine and Fisheries, requesting to advertise the time signal in the *Morning Chronicle*.[48] As Ashe pointed out,

> The present system seems to be most unsatisfactory, in which there is no advertisement whereby shipmasters coming to the Port can tell that there is a Time Ball and the hour at which it is dropped. This advertisement is a matter of secondary importance in harbours which are largely frequented by Steamships, as they, voyaging constantly to and fro between the same points, learn all necessary details in connection with the Ports visited. In the case of this Port, the vessels visiting which are largely sailing ships, and the Masters not of the most intelligent class perhaps, the usefulness of the time service is to some extent lost, on account of this lack of information.[49]

Mr Smith, a sceptical public servant in the Department of Marine and Fisheries, believed the *Chronicle* had placed the notice to convince the department to pay it $25 a year for the ad. As far as he was concerned, shipmasters were informed of the time ball directly by an agent, which was superior to a newspaper anyways.⁵⁰ An official sent to investigate reported back that, while the ad might be useful, he saw no reason why the time signal should be given special treatment over other matters important to shipmasters.⁵¹ That ended the correspondence until 1894, when the director of the Meteorological Service in Toronto instructed Ashe's successor at Quebec, Arthur Smith, to begin publishing newspaper notices when the time ball was wrong, along with the error, so that it could be corrected.⁵²

Amid these ongoing challenges, the General Time Convention of 1883, which prescribed time zones for U.S. and Canadian railways, threatened to reshape the Canadian timekeeping landscape. Immediately, railways (supported by observatories distributing time) switched en masse to standard time. But it remained unclear to what extent ordinary people would need to make use of standard time. The railways' decision evoked mixed reactions. Proponents such as Fleming were overjoyed. But for other people, particularly in rural areas, the change was hardly noticed. And for many it was not a change at all, as local time remained in use in places beyond major cities and railway stations. The Massey Manufacturing Company, makers of agricultural machinery, published an article in its circular, *Massey's Illustrated*, to explain standard time to rural customers a few months after standard time's adoption: "One of the events of the age is the arrangement of the 'Standard Time,' an achievement which this generation may well be proud of. We suppose our rural friends, who are not so tied to exact time, have not noticed the change like the town and city folks, who are not guided in their daily pursuits by the sunrises and sunsets but must move promptly to the minute."⁵³ As Massey indicated, standard time was largely an urban phenomenon.

While rural life went on unchanged, the cities faced more uncertainty. When the railways in Canada in mid-October 1883 first indicated their intention to change the time system, Toronto's *Globe* newspaper sent a reporter out to ask the opinion of dozens of merchants and business owners in that city and in Hamilton. The responses were mixed. One Toronto merchant, John Macdonald, thought standard time would prove "most inconvenient and cause annoyance to merchants and the travelling public. There seemed to be no pressing [need] for it, and the railways here might have continued on the present system, which the people now understand."⁵⁴ The president of the Toronto Board of Trade was similarly concerned about disruptions to business, but supported the change as long as it was permanent and universal; incomplete adoption would only make things worse:

The change would doubtless cause much inconvenience and annoyance to mercantile men, at least for a time. What should first be
ascertained was whether the system proposed would in all probability be a permanent one. If so, all the railway companies should come
into it. If one company kept outside it would prove awkward for the
travelling public, as there was 17 minutes difference between the proposed standard railway time and the local time. Not only should all
the railways adopt it, but the steamboats should follow suit. In fact,
there would have to be an uniform time for the city and railways …
If the proposal of some of the railway companies must go into effect,
all should join, so that no confusion would occur in going to any
train or steamboat. The experience of Port Hope showed the necessity for this.[55]

The comment about Port Hope was likely a reference to a sweeping
railway merger in 1882, which brought many small companies together
under the banner of the Midland Railroad. Timetables for all these small
rail companies had to be reorganized and unified – a messy process that
standard time was going to reproduce on an even grander scale across all
of North America north of the Rio Grande. It had to be done carefully if
confusion was to be avoided.

In Hamilton, most interviewees hoped for uniformity – they would
agree to the change as long as neighbouring communities adopted it
too. For example, the owner of Hamilton Bridge and Tool Works agreed
to the new railway time only if London and Toronto also adopted it,
while J.H. Stone, a cage and lantern manufacturer, said that, whatever
happened, he hoped Hamilton would use the same time as the rest of
the province.

Other Hamilton merchants were less concerned about uniformity
between neighbouring cities, and more concerned with keeping railway
time and civil time unified within Hamilton itself. Mr D. Moor, for example, was in favour of moving of Hamilton's civil time nineteen minutes
earlier to match the new railway time, because "he did not wish to see
two kinds of time in use in the city."[56] Mr Burrow of Burrow, Stewart, &
Milne, malleable iron work, also "did not want their office clock and railway time different. It would be a great inconvenience to their customers
coming to Hamilton by railway and probably be depending on railway
schedule time to leave the city and going to the station find himself left on
account of Hamilton time being some nineteen minutes slower."[57]

Other respondents confessed fears that railway time might inconvenience their employees. After all, adopting it meant starting work nineteen

minutes earlier than usual. For this reason, Mr Bell, secretary-treasurer of the Ontario Cotton Mills, did not want Hamilton to change its civil time to match railway time. His employees commenced work at 6:30 a.m., and losing nineteen minutes of sleep would be hard on them. The firm's manager, Mr Snow, was more relaxed: "If they found the new time a little early for their operatives they could change to suit, or have a mill time of their own, as is now the case in the Canada Cotton Mills, Cornwall, where the mill time and the town time are entirely different."[58]

Not every workplace could be so flexible. Still, it did not bother one local maker of boots and shoes, who thought starting work earlier might benefit his workers, because they could then go home earlier. The same man also opined that "the day's work of ten hours was better placed between daylight and dark on the proposed new time, and would consequently be a saving of gas, especially in the fall and spring, by the workmen quitting earlier in the evening than they now do."[59] R.M. Wanzer & Co., sewing-machine manufacturers, similarly felt that it would be "a little inconvenient for the men to get around nineteen minutes earlier for a while, but when all clocks and time was changed a while they would not know the difference," and that the change was "better for the workmen than the present time used, as they got away earlier in the evening from work."[60] In some industries, workers already made their own schedules anyways, like Mr Burrow's iron workers, who, as he told the reporter, "have always run their shops on a time of their own."[61] For them, the new time change was trivial.

In the end, Hamilton city council decided to adopt railway time as civil time, as did Toronto.[62] Smaller towns all along the rail lines southwestward towards Windsor and Sarnia followed suit. Sarnia, according to longitude, was actually too far west to be in the same time zone as Toronto and Ontario's other major urban centres, but political and business connections trumped arithmetical precision, and Sarnia adopted Toronto time. Andrew Gordon of the Marine and Fisheries Department noted that the only small inconvenience this might cause Sarnia's residents was that "banks in all places west of the standard meridian might, until standard time received the force of law, have to keep open after 3 o'clock for the payment of bills ... for the number of minutes by which the local time differed from the standard."[63] He felt that the benefits outweighed the costs of this arrangement.

Regardless of opinions, the railways implemented their plan across the United States and Canada at noon on Sunday 18 November 1883. Considering its scale, the transition went remarkably smoothly, but it was not without hiccoughs. On 21 November, for example, the *Globe* reported that in Boston "the first legal complication arising from the change of the time occurred today. Notice for the examination of a poor debtor was issued last week from the office of the Commissioner of Insolvency, returnable at 10 o'clock this morning. The insolvent appeared at 9:48 standard time, but the Commissioner ruled that it was after 10, and defaulted him. The case will probably go to the supreme court."[64]

Nothing so complicated was reported in Toronto, where standard time was widely adopted, though with exceptions. The High Court of Justice had ordered that its clock not be changed, despite all the other clocks in Osgoode Hall (where the court was seated) reading standard time.[65] According to the *Truth* newspaper, it took just a week or so for Torontonians to adapt to the time change: "There was some confusion Sunday last, but everything is now serene and most people don't know the difference, except that the mornings are a little longer dark after six o'clock, and the evenings vice versa."[66] The University of Toronto's student newspaper, the *Varsity*, offered light-hearted advice to student regarding the time change, joking that students might make the best of the change by spending the extra time with their sweethearts: "Undergraduates who have engagements for next Sunday evening to accompany young ladies to church, are reminded that it will be necessary to pass a given point 17', 34" earlier than formerly, as the Standard Time comes into operation at noon on that day. It is equally important, however, to remember that for the sake of old times they may leave the family residence after church at 17', 34" later than the apparent time on the parlor clock, which the thoughtful small brother will doubtless set on the new method."[67]

The actual process of changing clocks on that Sunday in November 1883 was clearly a trivial matter, easy to understand and carry out. What complicated the process was that other timekeeping practices did not immediately disappear. Standard time now existed alongside the other times, rather than replacing them entirely, leading to a lack of clarity for decades to follow A new urban/rural divide was created, and even within cities holdouts of the old time remained. Travellers continued to have to confirm which times their destinations used. In 1896, for example, one inquirer to the Quebec Observatory was still unsure whether the time gun at Quebec was fired according to standard or local time.[68] A decade of use had not yet entirely solidified the supremacy of standard time in Canada.

Between the rural residents who did not have much use for standard time and those urban dwellers who made the change more readily, there were some who were well aware of the time change, but vocally opposed it, finding it both confusing and frustrating. Fleming received a series of complaints from G.W. Wicksteed, who was the law clerk of the House of Commons. Wicksteed was not against the railways using standard time, but he preferred that local time continue to be used for daily life and for the purposes of governance. A few weeks after the railway change, he wrote a series of letters to Fleming expressing confusion about the new system:

> I do not think you intended that that time [railway time] should be legal time for regulating all the business of life, to which I do not think it can be made applicable ... [but] the people of Quebec, or a great many of them, believed that the Railway standard applies to all the affairs of life civil and criminal, and as I find that the Citadel Gun has been fired, and church services commenced by that Standard Time ... and that our parliament clock had been altered to it, I cannot but think that this impression is very extensively entertained and may lead to many very undesirable consequences, which you never intended.[69]

Wicksteed suggested that Fleming correct the public on this "misinterpretation." But Fleming, of course, was pleased that railway time was being widely used and would do no such thing. In subsequent letters, Wicksteed continued to try and convince him that there was a better way. The time "jumps," as Wicksteed called them, at the edges of each time zone, were highly problematic for people living near them.[70] He conceded that standard time might be useful for travel and for science, but wanted local time left alone. Or, better yet, use a single, universal time for science and railways around the world and local time for everything else.[71] Anything, he said, would be better than the time jumps.

Wicksteed kept up a polite but lively debate with Fleming for years. In 1885, a particularly lengthy missive complained about the time jumps at each line of longitude, which, he worried, might undermine the electoral process.[72] "We have no natural time zones, but five divided by invisible lines," wrote Wicksteed, comparing the situation in Canada to that in the United Kingdom, where the Irish Sea conveniently separated Dublin time from Greenwich time.[73] These arbitrary lines used in Canada, he continued, were "very hard to find (as you know) even by men of science ... And yet on one side of such lines it is late day, and on the other side by law night. A returning officer on our side of such a line in our county of Essex, would have to close his office an hour sooner than his deputy on the other side – and yet each of them are bound by law to close at the same hour, and a man may lose his election by having some of his votes received before or after the legal hour."[74]

Wicksteed was surely overstating the actual threat to democracy, but he was right to raise the question of time and the law. What time was legal time? There were situations where it mattered. A 1893 case in London, Ontario, for example, questioned the hours in which it was legal to serve alcohol.[75] Two bars had stayed open until 10 p.m. solar time, rather than standard time, earning themselves a half an hour extra serving time.[76] According to the *Ottawa Free Press*, the magistrate ruled that "all authorities were in favor of solar time, and without some act to legislate Standard Time the other must govern. He therefore dismissed the case. The effect of the judgement will be to allow bars to remain open nearly half an hour later Saturday nights, as well as every other week night."[77]

Wicksteed published some of his own concerns in the *Ottawa Citizen*, reiterating his worry about election fraud, but also about the legal implications of the time change. He wrote, "Our criminal law defines 'night' to be 'the time between nine o'clock in the evening and six o'clock in the morning on the next succeeding day.'"[78] As the time zones created one-hour jumps at each line, court cases would become complicated, because "the law draws very important distinctions between crimes committed in the night or in the daytime."[79] In the *Legal News* in 1885, Wicksteed explained that "Burglary" was defined specifically as a crime occurring in the night only. Similarly, insurance claims and mortgages might run into timing problems on either side of a meridian line.[80] Wicksteed was not alone in his concern. A contributor to the *Quebec Daily Mercury* wrote the following diatribe against the new railway time zones:

It seems, indeed, amusing that so absurd an innovation should be so quietly received, not only by our country cousins across the line [United States], but also by the community at large. Without any apparent participation of the Executive, without any proclamation, or valid sanction of any kind, it is, for a time at least, an established reality, solely by the will and pleasure of a few visionary dreamers, gifted, apparently with an immeasurable capacity for fantastic fancies and baseless speculation. I would not be mistaken. The railway time system, as suggested, is eminently fit and proper for railway purposes – but for these alone. The local time need not have been interfered with. Both systems might, as hitherto, have gone on concurrently. They manage these things otherwise beyond the sea. Even in so simple a matter as the determination of the first meridian, it is with manifold deliberation, with cautious counsel, and with the aid of men of unquestioned position in the world of science, that they advance towards a decision. We may not have authorities of the same calibre on this side of the Atlantic, but fortunately, the question here at issue is one on which any person of ordinary education is competent to judge, and on which even the least instructed may easily become informed. A brief experience of the anomalies and inconveniences attending the new system will probably suffice: and the solution may be safely left to the common sense of the people, a criterion which, in the long run, never fails.[81]

Several things stand out in the letter. Like Wicksteed, the author points out the impromptu nature of the time change, with no government sanction or legal framework. International law was just as lacking. The change preceded the IMC by a year, so there was not yet any international precedent for using Greenwich as the baseline for timekeeping. Indeed, the author was clearly aware of the European deliberations over the prime meridian, although his interpretation of them as "cautious counsel" is somewhat oversimplified. But this appeal to the authority of science, in opposition to "dreamers" like Fleming and his railway associates, paints standard time as intrinsically unscientific. This argument is a fascinating parallel to the Rome Conference and the IMC. At these conclaves, scientists wanted nothing to do with standard time – they sought a universal time for professionals alone – and did not intend to change the timekeeping habits of the public. This author shared that opinion: time reform for railway professionals and scientists was fine, but forcing the same on the public failed to pass the test of "common sense." What is more, it lacked legal authority.

Fleming recognized the weak legal foundation of his innovation, and made multiple attempts to rectify the problem, citing as precedent the British law from 1880, which had made Greenwich time legal time. But he was unable to secure a dominion law in Canada to affirm standard time. In one of his more concerted efforts to do so in 1891–92, he was thwarted by Major-General D.R. Cameron, son-in-law of his friend Sir Charles Tupper and head of the Royal Military College.[82] With the bill defeated, legal time in Canada would henceforth be determined by individual provinces, not by Ottawa.

Cameron's opposition to the bill is worth examining. There was an extensive correspondence between Cameron, Fleming, and the Department of Marine and Fisheries over the issue, but Cameron summarized his position in a letter to the department from November 1891:

1   Neighbours are most concerned in daily routine with sun time.
2   People remote from one another are interested mutually in relative time dissociated from sun time.
3   Zone time satisfies neither directly – i.e. without ever varying calculation.
4   Local time and Universal time immediately and directly satisfy all possible cases.[83]

Cameron, in other words, agreed with the scientists at the IMC. Universal time should be used for special purposes, and local time for everyday life. Trying to combine the two by using standard time complicated time-keeping for everyone and pleased no one.

Fleming retorted: "Zone time needs no defence. It is not an untried theory or experiment. It has been in daily use in Canada and throughout this continent for nearly nine years."[84] While Cameron may be right that many Canadians still used older methods of keeping time, surely the influence of the railways and telegraphs would soon make zone time universal. "I venture to think that the day is not far distant where no person will dream of using any other reckoning; unless perhaps those persons who are in out of the way places such as the remote ports of the Hudson Bay Company."[85]

Cameron melodramatically justified his objection to zone time as "an attempt to prevent a national calamity."[86] But his theatrics failed to convince Charles Carpmael, superintendent of the Dominion Metrological Service. Carpmael was one of the key figures in Canada's time service and in astronomy (he had also participated in the transit-of-Venus observations in 1882).[87] He supported Fleming's attempt to make a dominion legal time for Canada, despite Cameron's objection.[88]

The arguments used in the debate over this bill should seem familiar by now. Whether universal time or local time ought to be paramount, with standard time as a compromise, had been thoroughly flogged over and over again. No one seemed to object to the existence of all these types of time; the debate was about which one was authoritative, and to whom. The railways, telegraphs, and observatories could have all the specialized time notations they wanted, but the public, according to people like Cameron, ought to be let alone to use local solar time. In the short term, Cameron's view won out. But the unofficial influence of standard time, under the direction of the railways, was hard to ignore, and Fleming, in particular, made a concerted effort to sell the virtues of standard time to the public. Even after the IMC, standard time based on Greenwich needed to be lobbied for, advertised, and sold. Fleming worked hard to teach the public about it, and his main target, beyond lobbying Parliament to make standard time the law of the land, was children and schools.

TIMEKEEPING IN SCHOOLS

Having failed to establish a legal framework for standard time, Fleming hoped instead to solidify its primacy through the inculcation of social norms. In other words, if the public could not be coerced to use standard time, it had to be taught to use it. A massive public-education campaign was required.

Public schools were the logical instrument of dissemination. Fleming wrote a pamphlet for the Canadian Institute in 1888 meant to explain time zones to children, which he intended as a teaching tool for schools, titled "Time and its Notation."[89] Fleming hoped the pamphlet would be used in the United States as well. In fact, he had first written it in 1887 for the American Metrological Society (AMS), but that society rejected it. The AMS members subjected it to some harsh criticism. Its biggest problem, some said, was that it was not well-written for an elementary-school setting and lacked the clarity needed to be a proper pedagogical tool. Barnard suggested making it briefer and clearer, so that teachers and children might get more out of it. He also pointed out that Fleming failed to include practical examples as teaching aids.[90] William Allen had similar concerns: "I fear it would not be utilized by the average teacher, nor be understood by the average pupil. If it were possible to condense the ideas expressed, in about one-third of the space ... then I am inclined to think the attention of both teachers and pupils could be secured."[91] Fleming was not used to writing for such a young audience.

Instead of rewriting the pamphlet, Fleming simply turned to a different society to publish it. The Canadian Institute accepted it and forwarded it to the governor general, with the intention that it be not only distributed to Canadian schoolteachers, but "transmitted to the governments of all foreign countries for the information and use of their educational authorities ... [as well as to] the Minister or Superintendents of Education in all British Colonies and Possessions."[92] The governor general, the Marquess of Lansdowne, who was nearing the end of his term and was on his way back to London, promised to carry out its request in person, making its case to the government in London.[93] Copies were sent to the Netherlands, and most other countries with which Britain had diplomatic relations, except Bolivia and Venezuela, which had no British embassies (Venezuela and Britain had cut off diplomatic relations in 1887 due to a territorial dispute concerning British Guiana, and the British embassy in Bolivia had been replaced with a joint embassy in Peru). The copies intended for the Orange Free State had to be forwarded indirectly via the Cape Colony government.[94] After they were received, the acting president replied that "our state being wholly surrounded by other colonies and states, no other system of universal time notation can be established than that adopted by our neighbours."[95] Italy and India also received the pamphlet, while Hong Kong asked for twenty-five extra copies.[96]

The Canadian Institute also distributed the pamphlet domestically. One of the most favourable responses came from the superintendent of the Ontario Department of Education, who asked for five hundred copies "for distribution to the inspectors of public schools and the head-masters of high schools throughout the Province."[97] The New Brunswick school board ordered three hundred copies, while the school board of the North West Territories ordered two hundred.[98] The Catholic school board of Manitoba was less supportive, but did promise to put it before the board's next meeting.[99] The superintendent of the province's Protestant school board meanwhile wrote back that "as the new time notation is in almost universal use in this province, there will be little difficulty in having it understood ... It will not be long before the new system becomes universal."[100] The other provinces did not immediately request further copies, but did distribute the few that the Canadian Institute sent up front for review by boards and teachers.[101]

Other proponents of standard time more adept than Fleming at pedagogy created teaching tools about the new time zones. One Philadelphia publisher sold a "time chart" for schools and mailed it to Fleming in hopes of expanding his customer base to Canadian schools.[102] This chart was much more accessible than Fleming's own pamphlet. It was designed with teachers, students, and the learning process in mind. The advertisement for it included reviews from teachers, which, though obviously curated to include only positive reviews, showed how to use the chart's visual aid in classrooms. The visual aid simplified the subject in a way Fleming's pamphlet did not. As one teacher wrote, "Longitude and Time is the most difficult section of the arithmetic to explain," but the time chart made it easy.[103] Another teacher wrote, "By it [the chart], dull pupils understand at a glance what attempted explanations failed to do."[104] Another, expounding on the difficulty of teaching time zones, found it "a valuable invention, throwing much light on a dark subject."[105]

Maps were already common in schools, and as early as January 1885 maps with standard-time zones were available to teachers.[106] In the mid-1880s, the *Home and School Supplement,* an illustrated monthly magazine edited by Seymour Eaton, included information on teaching standard time in several issues. The September 1885 issue included a page of exercises, asking students to find the time difference between major cities, or how standard time differed from local time.[107] The March 1886 issue included an ad for a standard-time map.[108] By decade's end, educators were not lacking for materials.

Although standard time seemed to be entering the curriculum fairly widely in Canada, Fleming wanted more. He sought support from his U.S. peers through the American Society of Civil Engineers (ASCE), giving a report to its annual meeting in 1888 regarding the teaching of time in schools. Quoting the Manitoban education official who had said that use of the new time would surely soon be universal, Fleming argued that teaching it to the next generation was the best way to secure its continued use. Fleming always associated standard time with twenty-four-hour notation, and he conflated the two in this report. Where the Manitoban letter had stated that standard time was used by almost everyone, Fleming took that to mean twenty-four-hour time as well. But that was hardly the case. According to one Manitoban railway worker, the twenty-four-hour clock, which Fleming so closely associated with time zones, was confined to villages, while solar time, measured in twelve-hour increments, continued to be common among farmer.[109] Apparently, when given the time in twenty-four-hour notation, "those from the country do not know what to make out of it ... The question is generally asked 'what is that by our old time?'"[110]

Nonetheless, Fleming was sure that teaching twenty-four-hour notation to children would bring about reform even in rural areas, as these young people would "carry them to their homes, and thus educate their fathers and mothers ... Through the medium of schools it is believed that in a comparatively few years the people will have their minds familiarized with the whole question."[111] Fleming suggested the ASCE forward copies of his pamphlet to the American Bureau of Education and press it to "enjoin teachers to give special lectures and lessons on the subject to the pupils attending the schools."[112] One committee member, Fred Brooks, called Fleming's proposal "beyond the province of the society," but the rest of the committee took Fleming's side over Brooks.[113] In any case, by the late 1880s standard time was being integrated into school curricula across Canada and the United States, though in spite of Fleming's poor pedagogical prose, rather than because of it.

INDIGENOUS USES OF STANDARD TIME

It is difficult to ascertain how the students on the receiving end of
Fleming's pamphlet felt, or what they took away from their classroom
lessons on time. What is available suggests a somewhat mediocre outcome, particularly in rural areas. This is not to suggest that students did
not grasp standard time. Rather they simply had little use for it. Yet rural
schools taught the concepts, and schools for Indigenous children were
no exception. This section discusses the reception of standard time in
schools in two Indigenous communities in southern Ontario. The experience of teachers and students in these schools, and the impact on their
communities, offer an important contrasting perspective on the reception of standard time. They demonstrate that the vision of modernity
embraced by standard-time advocates was malleable and open to various definitions and uses. Indigenous communities embraced or rejected
standard time as it was useful to them, in order to navigate the specific
challenges and prejudices they faced.

The schooling of Indigenous children in Canada is caught up in the
history of colonial oppression. As historian Jo-Ann Archibald writes,
"The beginning of institutional schooling signalled the beginning of the
decimation of many First Nations' societies ... Even though the children
were provided with an education which ostensibly was to enable them
to 'fit' in to mainstream society, the truth was that they did not."[114] The
young people were taught to assimilate, but prejudice prevented their
full integration as equals. The schools themselves were often oppressive
spaces. As J.R. Miller writes, residential schools in particular were "an
instrument of attempted cultural genocide."[115] In many places the school
system "severed the ties that bound Native children to their families
and communities, leaving semi-assimilated young people and shattered
communities."[116] These schools had a "hidden curriculum" behind the
subjects of arithmetic and language, and this was assimilation.[117] Under
these circumstances, standard time and clock time became tools of colonialism. School inspectors measured students' adherence to clock time
and considered punctuality a sign of progress towards assimilation.[118]

Some Indigenous peoples, like the Mississaugas of the Credit First
Nation, managed to stave off the worst of this assault, keeping their
identity while associating uneasily with local settler communities,
using the school system to their advantage whenever possible. In the
mid-nineteenth century, Peter Jones (Kahkewaquonaby), a prominent
Mississauga chief, Methodist, and promoter of education – he was a
friend of Egerton Ryerson – wanted "the schools eventually to be run

by Indians, to produce duplicates of himself: men and women able to compete with the white people, able to defend their rights in English, under English law."[119] His descendants continued to adapt the system to their advantage as best they could, in spite of prejudice. Accordingly, the Mississaugas of the Credit First Nation developed a deep interest in education.

By 1884, the community had developed strict rules for its schools. Teachers enforced attendance with both the carrot and the stick, suspending pupils for truancy, but also rewarding students with the best attendance and parents of regular school attendees.[120] Classes lasted from 9 a.m. to 4 p.m., with an hour for lunch and two recesses of fifteen minutes. Tardiness and early departures were carefully regulated, as was the ringing of the schoolbell. The teacher was required to ensure that "the school house should be ready for the reception of pupils at least 15 minutes before 9 o'clock a.m."[121] Most significantly for our purposes, the rules and regulations for 1884 stated that "the time to be used in school work shall be the 'New Time' of the 75th Meridian."[122] Only a year after standard time entered North American railways, the schools of the Mississaugas had eliminated competing time systems from their education.

Schools on the Moraviantown reserve in southern Ontario took up standard time just as fast, leading to a heated debate among the Lunaapeew people of the Delaware Nation concerning their schools and timekeeping on the reserve in 1886. Moraviantown was home to the "mission school" and the "reserve school." The former stood next to the River Thames and was funded privately by the Moravian Mission, before eventually being sold to the Methodist church.[123] Between 1885 and 1898, the teacher was Dora Miller, an English woman whose salary was paid by the Moravian missionary society. The reserve school, in contrast, went through at least six teachers between 1885 and 1898.

Time measurement in Moraviantown, like most places in the 1880s, was in flux. Local solar time was the simplest method, but the Lunaapeew at Moraviantown were certainly aware of the railways' adoption of standard time in 1883. The reserve school had its own bell in a belfry on the roof, which was rung regularly on schooldays.[124] In fact, it was rung so often that its vibrations were apparently causing structural damage, and an inspector in 1893 recommended moving it to its own framework separate from the schoolhouse.[125]

But the bell was not for the students alone. It was also "rung to mark time for the neighbourhood."[126] The local-time debate in Moraviantown centred around how the teacher determined the time for that bell. The teacher in 1886 was Daniel Edwards, who had taught there since 1877. But members of the Moraviantown Council wanted to replace him. As the school inspector reported:

> There is a desire on the part of the Indian council to remove Mr. Edwards, and employ in his place Mr. James Stonefish, a young Indian who has returned from Nazareth, Penn. I could not recommend that Mr. Edwards be thus summarily dismissed, but as he intends to resign at the close of this year, I advised Mr. Stonefish to try meantime to pass at least H.S. Entrance Exam and to attend the Model School. I should be glad to see Mr. Stonefish in charge of the school after Mr. Edwards resigns but as he has had no training I do not consider him at all equal to the present teacher. I believe that Mr. Stonefish has a good English Education.[127]

The issue of replacing Edwards was controversial, and it played out publicly in letters to the editor of the *Indian,* a newspaper based in Hagersville, published by Dr Peter Edmund Jones (the son of Peter Jones mentioned above).[128] The *Indian,* which ran for twenty-four bi-weekly issues in 1885 and 1886, was, according to its tagline, "a paper devoted to the Aborigines of North America and especially to the Indians of Canada."[129] A correspondence section printed local news and opinion pieces from Indigenous communities around Ontario.

When the Moraviantown Council moved to replace Edwards, it also proposed to ask the dominion Department of Indian Affairs to be allowed to appoint its own school trustees and teachers, claiming that inspectors and teachers had been lax in their duties.[130] Its specific complaint against Edwards was that he had been arriving late to school, especially in the winter. But not everyone was pleased with the Council's chosen replacement. As we saw in the inspector's report, Stonefish did not have the same qualifications as Edwards. An anonymous correspondent, who signed with the initial "W," wrote to the *Indian* explaining that they did not care who taught at the school, as long as the teacher was qualified. Stonefish, according to W, was not qualified. Furthermore, W claimed, the charges of Edwards's tardiness were unfounded. The members of the Council, W wrote, "complain of the teachers arriving late in the morning during winter, yet there is no clock in the school house nor any standard time on the reserve."[131] W, keen to undermine the Council's charges against Edwards, invoked the ambiguous nature of timekeeping on the reserve to exonerate Edwards. W then wrote to Lawrence Vankoughnet, the dominion superintendent of Indian Affairs, to complain about the Council's actions.

A response to W followed in the early May issue of the newspaper, written by John Noah, one of the Council members who had voted to replace Edwards: "The clever writer of the said article [W] talks at random, irrespective of telling the real truths and facts. Just fancy a man of common sense saying that there is no standard time in Moraviantown? Time, we have the Hamilton railway time regular from Bothwell; we even hear the town hall bell every day, and all our clocks and watches are set accordingly; and this clever man says there is no standard time."[132] The Council, he reported, passed the motion "in order to encourage, as much as possible, our own young men of education and ability to devote their time and talents for the good of our fellow Indians."[133] Noah then defended Stonefish, saying that he had received his teaching certificate in Pennsylvania and was perfectly qualified.

Standard time, and access to it, were being used as the central argument over who had the right to appoint teachers. W suggested that time on the reserve, as regulated by the school bell, was unreliable, and therefore calling Edwards late was impossible, given that he was quite literally the arbiter of time for the community, since he was the one who rang the bell. However, Noah, a representative of the Council, claimed that the reserve did indeed have access to standard time and did not rely on its schoolteacher for the time, but rather could hear the time signal from the nearby town of Bothwell. Noah, in his attempt to claim some autonomy for his community, relied on the prestige and legitimacy that came with access to the new, modern, accurate standard time to do so.

But W was not the only critic of the decision to hire Stonefish. A third letter to the *Indian*, published in July, refuted Noah's claim that the reserve had easy access to standard time. The author was James Dolson, a thirty-year-old Lunaapeew man, who had been married to Johannah Hill by the Rev. A. Hartman of the Moravian mission two years earlier, in 1884.[134] Dolson wrote:

> Mr. Noah requests you to "fancy a man of common sense saying there is no standard time in Moraviantown," and I request you to fancy there is standard time in Moraviantown ... For, assuredly, it would be but fancy, as the capacity of the Bothwell bell is but two miles, and the *nearest point* in the reserve is three miles, and the central point [where the reserve school was located] four and a half miles thence; hence the bell cannot be heard by us for a month at a time sometimes; only when the weather and wind are favorable, which is but seldom. Therefore the charge against our present

teacher, Mr. Edwards, that he does not always call school, sharp, at 9 a.m., is merely a supposition, a catcall of his enemies.[135]

Dolson's measurements to Bothwell are reasonably accurate, but the range of the bell is harder to ascertain, making it difficult to know whether Noah or Dolson was correct. But in this case, knowing who was right is not necessary to be able to draw some conclusions about the whole affair. What is important is that people on the reserve were aware of standard time and considered it to be authoritative, and thus Noah's assertion that he had access to standard time was also a claim to share its authority. Noah, in his desire to achieve some autonomy for the Council and the community, appealed to the authority of standard time based on Greenwich, claiming it for their own.

Edwards ultimately did leave his position as teacher of the reserve school, but he was not replaced by Stonefish. Stonefish went on to pass his high-school entrance exam and attended Ridgetown Collegiate Institute instead.[136] After his departure, Edwards was followed by a revolving door of new teachers in quick succession. Like elsewhere in rural Canada and across the globe, standard time and local time would continue to exist awkwardly side by side in Moraviantown. The teachers who succeeded Edwards would continue to act as the informal time-keepers for Moraviantown, in uneasy competition with the distant toll of standard time from the Bothwell bell.

CONCLUSIONS

As is apparent from the varying receptiveness of provincial boards of education, the proliferation of pedagogical supplements, and the time-keeping controversies in Indigenous schools, Fleming's dabbling in educational curricula produced mixed results. While it raised awareness about standard time, it could not enforce its use. People, regardless of nationality, class, race, or gender, used the time that was convenient to them. Many ignored standard time or treated it as a joke. But people made use of standard time for their own purposes. It could be, and was, a status symbol, allowing its users to appear "modern." The Council of the Delaware Nation at Moraviantown used it to make claims about political legitimacy, in opposition to colonial school inspectors, who similarly used clocks and schedules as a measure of how "westernized" a school was. This tactic is not dissimilar to that of St John Winne in Britain, who advertised his electric time service as modern, condemning Ruth and Maria Belville as unmodern, comparing a masculine, professional modernity with a "strange," outdated femininity. Controlling the narrative about timekeeping could be a useful tool for diverse purposes. In the United States, conflict over the place of government oversight in timekeeping, and the right of universities to profit from their distribution of accurate time, reflect a similar attempt to define the nature of timekeeping and control its dissemination.

The thread that unites all these cases concerning standard time is that they tell a story about power relations and the construction of authority. The privileged delegates to the IMC, regardless of nationality, made their decisions in a realm largely disconnected from ordinary people. Indeed, as we saw, most of them were concerned only with scientific timekeeping. They had no intention of altering public clocks. Yet the decisions they made had ramifications beyond their expectations. Although Greenwich time was not meant for use by everyone, the association of Greenwich time with authority and "truth" made it desirable, as access to it became a trapping of status or profession. Helped along by promoters like Fleming and Allen, standard time, as a subsidiary form of Greenwich time, became accepted as the ideal form, the categorical imperative, of time measurement for the use of business and travel. But access to it was limited, and its practical application to rural life was negligible, so its implementation was slow. Meanwhile, other forms of timekeeping did not just disappear to make room for it. Standard time did not simplify timekeeping by sweeping away the competition, but rather added an additional temporal standard to operate alongside those already in place.

Educational institutions helped imbue standard time with its perceived authority. This occurred directly, as universities were the distributors of standard time throughout the United States, but also indirectly, through the curriculum imposed on students across both countries. Where reformers failed to enforce "modern" timekeeping by law, they espoused it through education. School schedules and strict truancy laws enforced a "modern" time-sense, which shaped perceptions of both class and race, as non-white and poor children were painted with the moral failings associated with lack of adherence to modern clock-time. Yet modern timekeeping could be claimed by marginalized people as well in order to assert their own modernity, as the Council did at Moraviantown. Education and institutional timekeeping were colonial tools, but they could be subverted, repurposed, and challenged. These contradictions are indicative of the larger debate about timekeeping at the end of the nineteenth century. Greenwich time did not instantly become the basis for a universal, undisputed system of timekeeping like Fleming imagined.

Nor would it remain the esoteric professional tool that scientists at the IMC envisioned. Instead, modern timekeeping would be an uneasy compromise between the two, warped by social and cultural contexts – a shifting standard, as messy, diverse, contested, and complex as the peoples and communities that used it.

Conclusion

This book began by tracing the history of an idea – a global, standardized timekeeping system. That idea had no single point of origin. It was conceived independently by various nineteenth-century thinkers. One of those people (not the first) was Sandford Fleming, a railway engineer, who spent a great deal of time and energy promoting the idea. Fleming did not set out to change the world with this idea, at least not at first. He was a professional engineer, troubleshooting a mild, but industry-wide inconvenience. He wanted to simplify rail timetables, which were inefficient because they needed to incorporate dozens of local time systems. Fleming's idea was to reduce the number of local times to a manageable number – just twenty-four for the entire globe.

The railway industry in North America liked the idea. Under the direction of William Allen, these new time zones were introduced in 1883. It was a quick and painless change in most major centres, and quite a few municipalities even decided to change their local time accordingly, although others preferred to continue using local time, converting to rail time only when necessary for travel.

Fleming, involved as he was in diverse transnational projects, was not satisfied with standardizing the timekeeping of just one industry on just one continent. He was a global citizen, as much as it was possible to be so in the nineteenth century, and an imperialist – those two identities were not mutually exclusive in his mind. He wanted to extend his idea in width – spreading it worldwide – and in depth – extending it to other industries and to other facets of civil life. Although he was solving a railway problem, he believed his solution had applications in other industries.

To succeed, standard time required participation on the part of members of the travelling public. At the very least, they had to understand how to convert their local time to railway time when they wanted to travel. But would it not be even simpler, Fleming suggested, if everyone just switched to railway time for all aspects of their lives? Here is where Fleming's simple professional troubleshooting evolved into a potential paradigm shift. One system of timekeeping, for everyone, everywhere.

In a different profession on a different continent, another idea was stirring. Among European geographers and astronomers, there were plans in the works for a single, universal time for the globe. This was a smaller idea than it sounds. It was not intended that everyone would use it – only astronomers, geographers, and navigators. Local time would remain the norm for civil life. A universal time would simply mean that astronomers from different parts of the globe, measuring extraterrestrial phenomena like the transit of Venus, could measure the same phenomenon using an identical notation. They too were engaged in professional troubleshooting.

Both ideas began small. Railway standard time and astronomical universal time could have existed side by side, without friction, without ever acknowledging each other. But that was not to be. These two separate ideas, dreamt up by professionals with their own separate networks and their own spheres of expertise, collided. That collision has formed the nucleus of this story. It is a case study of how knowledge about – and authority over – time were constructed out of conflicting ideas. Like most scientific discoveries, modern timekeeping was not objectively "discovered," or "observed," or "invented." It was produced historically in a "decision-laden" process, to borrow a phrase from Karin Knorr-Cetina.[1] Individuals, working within their specific cultural contexts – not least of which included their professions – shaped modern timekeeping practices. These contexts both limited their scope of action and allowed them to manipulate cultural norms to cement their ideas.

Understanding it this way, one can easily see that the central conflict at the IMC in 1884 was between professions, not nations. The two timekeeping proposals had little to do with each other, and neither side was keen on the ideas of the other. Indeed, astronomers did not seek sweeping time changes at all, focusing instead on a prime meridian for determining longitude for navigation, rather than for civil timekeeping.

Astronomers' concerns at the IMC were products of their own context. Britain's Science and Art Department had chosen conference delegates based on their opinions on the metric system, not on their timekeeping expertise, because the metric controversy was the current topical crisis in the British scientific community in 1884. And the metric controversy, as demonstrated by the cases of Annie Russell, William Parker Snow, and Charles Piazzi Smyth, was itself embedded in a larger cultural context of religious belief and the attempt to use science to confirm scripture via archaeology, astronomy, mathematics, and other natural sciences. This in turn was related to a broad context of imperial competition, notions of racial hierarchies, and nationalism. All these factors, alongside the changing boundaries between amateurs and professionals, and various methods of exclusion, inclusion, and legitimation, almost blocked Fleming from attending the IMC. These influences also meant that the British astronomers at the conference repeatedly undermined Fleming, a fellow British delegate, because he had vastly different goals based on an entirely different set of core assumptions formed in a very different context. The astronomers were wary of making any changes to civil timekeeping. They felt that dictating the way the public measured time was well beyond the scope of the meeting and overstepping their mandate. To them, Fleming was a radical. To Fleming, they were short-sighted.

Standard time was soundly defeated at the IMC. But after the extensive lobbying campaign Fleming had carried out to promote it, much of the public conflated his idea with the conclave's results. The timekeeping ideas of two different professions had inadvertently become combined. Newspapers confused the universal time for astronomers with Fleming's plan to reshape civil time for the public. No one seemed sure anymore whether the IMC had sanctioned universal time for specialized tasks or for everyone. Watchmakers began applying in droves for patents that accommodated another of Fleming's proposals, the twenty-four-hour clock, which had also been grafted onto the conference's universal-time resolution.

The conflation of the two very different timekeeping ideas meant that the core assumptions of one idea were applied to the other and vice versa, adding new layers of complexity. The best example of this is the concept of accuracy. Unlike Fleming's standard railway time, astronomer's universal time required consummate attention to accuracy, down to a fraction of a second, to be of any use for the task for which it was intended. So when the public got the idea that universal time was to be used by everyone, some notion of heightened accuracy was parcelled with it. New commercial ventures latched onto this idea, attempting to sell accurate universal time, with variable success, to people and industries that had previously had no need for such temporal precision. These businesses were selling the future – a particular, technologically utopian future. With temporal accuracy came the promise of progress. Time-distribution companies offered the wealthy and the enterprising an alluring chance to associate oneself with cutting-edge modernity, and all of its trappings. Meanwhile, other entrepreneurs like the Belvilles marketed an alternative modernity, one with wider boundaries of social inclusion, yet still tied to the legitimacy of scientific accuracy produced by professional expertise and sanctioned by international diplomatic consensus.

The conflation of the two ideas also meant that neither side was particularly happy about the outcome. A frustrated and overworked astronomer royal at Greenwich, whose employees produced the accurate time that was sold to civilians, threatened to cancel the public time service, claiming it was extraneous to the observatory's purpose. He still clung to the original vision of universal time – that it was for professional use by astronomers and the Royal Navy, not for commodification or for public use.

On the other side, Fleming was frustrated by the scaling back of his time scheme. The IMC agreement did not include any synchronized form of civil timekeeping, let alone standard time, and he had a difficult time making it law even in North America, where it had found the widest acceptance. Neither Canada nor the United States enshrined standard time in law, while Britain legally adopted Greenwich time as a national time, but not standard time writ large. In spite of all the hubbub surrounding the IMC, local time continued to be the norm in much of the world.

Having failed in the legislatures, reformers instead tried to convince the public to switch to the new time not by law, but by education, creating curriculum aids for public schools. Yet education as a means of altering norms of behaviour met with mixed results. People continued to use the time that was convenient to them. In some cases, like the customers of the Standard Time Company in Britain, or the Council of the Delaware at Moraviantown, people claimed ownership of standard time in order to reflect its authority and modernity onto themselves for their own purposes. But others were just as happy to continue using local time and did so for decades. The unintentional association of standard time with astronomical time's excessive accuracy meant that standard time was not more broadly adopted until wireless-radio technology made the dissemination of accurate time more affordable in the 1920s. By that time, because the IMC had failed to institutionalize standard time as an international norm, the nation-state had become the principal determinant of timekeeping practices, only loosely following longitudinal time zones. Throughout the twentieth century nation-states proved themselves stubbornly resilient in the face of transnational forces.[2] Yet it would be a mistake to suggest that they alone constructed global timekeeping practices. The nation was not the source of legitimacy that established modern timekeeping as it now exists (see Figure c.1). As

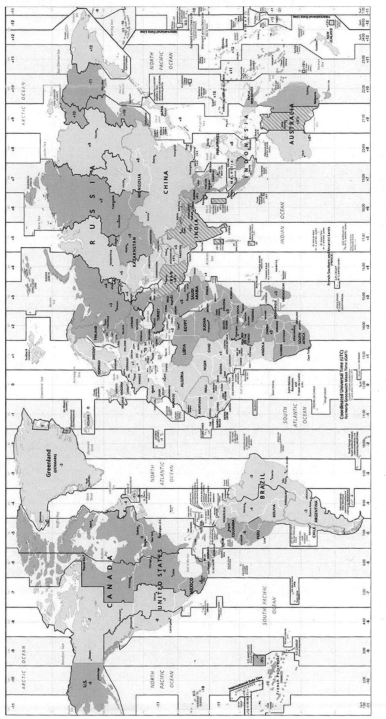

Figure C.1 | Standard Time map, 2020. Time zones follow political boundaries as often as longitudinal ones. China, for example, uses a single time despite crossing multiple zones.

this book has shown, professional and individual identity, wrapped in a cultural context (which included nationality, admittedly, but not exclusively), made time take the shape that it did.

As for the long-term legacy of the IMC itself, while the conference had almost no direct or indirect ramifications for civil timekeeping, its legacy in the diplomatic world was much more significant, because it set a long-lasting organizational precedent for the international coordination of timekeeping. Gatherings of technical specialists continue to coordinate and define international standards to the present day, and, just as at the IMC, these specialists are considered diplomats as well as physical scientists and engineers. In the twenty-first century, the topic of these gatherings has changed somewhat, as ground-based atomic clocks have replaced astronomical observations as the basis for timekeeping. Now, for example, scientists and engineers meet to regulate navigation satellite systems, which depend on coordinated atomic clocks in a variety of globally scattered laboratories. As another example, the length of the second is determined internationally through a pooling of national suites of atomic clocks at the International Bureau of Weights and Measures outside Paris. The IMC helped pave the way for these modern efforts at global coordination and standardization and, in this respect, cast a long shadow.

Of course, diplomatic convention is only half the story. The various reinterpretations of the IMC's resolutions by the general public equally shaped the discourse surrounding timekeeping. This dual source of legitimacy – expertise and diplomacy on one hand and public usage on the other – hints at a formula for establishing global norms that is generalizable to any field. Official diplomacy and professional expertise matter, but so do public acceptance, interpretation, and usage. A global norm might have an official definition enshrined by a meeting like the IMC, but its local adoption is often a symphony of variations on a theme, adapted to local conditions and local limitations. No global norm survives contact with the real world intact. Timekeeping, certainly, is no exception.

If there is a single, influential work that initiated scholarly discussions about timekeeping in history, it would have to be Marxist historian E.P. Thompson's 1967 article "Time, Work-Discipline and Industrial Capitalism."[3] Thompson's contribution has not yet been discussed in this book, as it focuses on the eighteenth century, not the nineteenth. Yet questions he raised are at the root of much that has been discussed, if indirectly, so it is worth delving into, and makes a fitting coda to this story.

Thompson's article is concerned with changing notions of time during the early industrial revolution. For Thompson, this change in "time-sense" was the direct result of technological improvements in the measurement of time and of the practices of the newly industrialized economy. He paints a stark division between before and after: for most of human history, time had been measured by nature. Fishermen and sailors scheduled their days around the tides, and farmers had to harvest during the correct season. Under natural time, work was task-oriented, and the day was organized around what needed to be done.[4] Work and leisure were not separate from each other. As Thompson put it, under natural time "social intercourse and labour are intermingled – the working day lengthens or contracts according to the task."[5] In the early industrial environment of the eighteenth century, however, natural time gave way to a strictly regimented clock time. There was, according to Thompson, an air of superiority about clock time. "To men accustomed to labour timed by the clock," he wrote, "this [natural, task-oriented] attitude to labour appears to be wasteful and lacking in urgency."[6] With the advent of accurate timepieces, employers were able to pay for their employees' time: they were paying no longer for the completion of a task but for a certain number of hours – hours that they did not want to see wasted. Time became currency.[7] Once established, clock time was not just enforced by employers, but inculcated by social institutions, including schools, which taught punctuality and industriousness, which Thompson calls "time-thrift."[8] Time measurement itself, he concludes profoundly, became an instrument for labour exploitation.[9]

Thompson presents a powerful demonstration of the effect of time measurement on human behaviour. His argument is largely convincing: Britain's industrial society changed its workers' very way of thinking about time, their "time-sense," in order to contract them into working regular, regimented hours. The clock forced them into "unnatural" patterns of labour by which they could more easily be exploited, as skilled work was replaced with poorly paid, unskilled positions. Thompson has been heavily influential, with good reason, but there are problems with his argument. First, his division between natural and clock time is too rigid. It paints the era of natural time as a golden utopia, evoking nostalgia for a simpler life when time neither needed to be nor could be understood "properly." But time measurement by nature is still time measurement. The change in time-sense required to live by clock time may not be so profound as he believes. As Paul Glennie and Nigel Thrift note, "Not owning a timepiece meant neither 'lacking information' nor 'lacking ability' to reckon with time."[10] Clock time mattered, and was understood, well before the industrial revolution. Second, Thompson's argument is entirely technologically and economically determinist. To a point, it is justifiably so. The timed workday would not be possible without accurate clocks, nor necessary without an industrial economy. But technological improvement did not create the demand for time measurement. Quite the opposite: the desire for accurate time measurement led to the development of the technology to do so.[11] In this way, Thompson has reversed the impetus and the result.

Nonetheless, it is hard to argue with Thompson's observation that eighteenth-century industrialization increased the use of clock time by the working class. The question becomes how well did those workers grasp time in the abstract, as distinct from "natural" time? Historian Vanessa Ogle has recently suggested that they did not really understand it. She shifts the moment of change in time-sense from the eighteenth to the twentieth century. She argues that, early in the twentieth century, many people could still not disassociate time from nature, and "strained to think of abstract time."[12] In studying the attempt to introduce summer time (daylight saving time) to Britain, Ogle suggests that many people were baffled by the notion that the time could be determined by law rather than by the sun. They were unable or unwilling to grasp time in the abstract, and thus their modern time-sense was perhaps not so developed as Thompson supposed.

I think there is a way to bridge the gap between Thompson and Ogle. Eighteenth-century industrialization may indeed have caused a shift in time-sense, as Thompson suggests, in which workers moved from task-oriented labour to schedules driven by clock time. I speculate that, given the events discussed in this book, late-nineteenth-century workers lived through a period of similarly disorienting change. By then, workers were used to clock time. But they had the rug pulled out from under them again when they were told that their local clocks, which they had long become accustomed to, were now lying to them. There was now a truer time – a perfect, universal, accurate standard – to which all other clocks must agree.[13] Yet they could not reach it, not easily. Greenwich time was costly and inaccessible to large swaths of the population. If Thompson's new time-sense in the eighteenth century was a means of exploitation, universal standard time might have been a new source of inequality in the late nineteenth century. Imagine having something so mundane as your ability to know the time taken away. The populace now had to either trust experts to do what before they could do on their own, or reject modern timekeeping altogether, as many did. It is no surprise that Ogle came across accounts of baffled twentieth-century persons crying foul at government officials – time experts – who were meddling again with the hours of the day. Sunrise and sunset were at least reliable, and free. No wonder natural, local time seemed a better option.

But trains cannot run by the sun. Expertise is necessary. This book is not meant as a polemic against experts and expertise in the modern world – today, the dangers of such anti-intellectualism are painfully obvious. But it is meant to demonstrate the ways in which expertise has always been tied up in social and economic privilege and to highlight the value of equitable access to information.

In the Introduction, I made two claims: that temporal knowledge was shaped, first, by individual agency and professional identity, and second, by debates about the nature of knowledge. We have now seen the process by which that worked. Astronomers claimed that accurate temporal knowledge was a tool for experts, not for use by the public. But their idea became tangled up with Fleming's notion of a universal civil time-keeping system for everyone, which had gained widespread, though not universal popularity. The result was a hybrid idea, combining the technical precision of the astronomers with the universality of standard time. Astronomers' professional and diplomatic status as experts and authority figures lent that new hybrid idea legitimacy and seemed to render, in theory though not in practice, all other sources of time obsolete. This created a demand for access to accurate standard time, and entrepreneurial capitalists like St John Winne and Maria Belville stepped up to fill that desire, for a price, leading to unequal access to the new temporal knowledge. Many people did not care and continued to use the time that was convenient to them. But unequal access led to a debate about the nature of Greenwich time – should it be a commodity, or a tool for experts, or a public good? Fleming advocated for the last, though not necessarily out of goodwill – he was as much a self-interested capitalist as Winne. Free, accurate knowledge of time would benefit his industry (railways), so he advocated for implementing it across all of society, attempting to use public schools to educate youth in its use. Time as a public good eventually won, but only after wireless technology allowed for it to be disseminated cheaply, undercutting the profits of those who sold time directly, like Winne. Inexpensive time distribution eventually aided other industries further up the chain – radio, aviation, and television.

The story of standard time has lessons that can be applied broadly, with implications for the modern, globalizing (I use both adjectives hesitantly) world. While this book is about the construction of temporal knowledge specifically, it could be about the production of knowledge in any form and the determinants of legitimacy. Individual agency and concepts of professionalism shape the production of knowledge, as do debates about the nature of knowledge: is it a public good, a commodity, or a specialized tool? The way that question is answered has vast implications. There is a case to be made about these patterns and processes applying elsewhere. The evolution of the internet, for example, might be understood in a similar manner to standard time. At its origin, it was used by experts alone, and was prohibitively expensive, but was sold to wealthy institutions. As it became more affordable – more than 60 per cent of the world's people now have internet access – new industries and new technologies were built on top of it. It too was not the result of one inventor but rather was constructed out of the interaction between multiple ideas from various contexts. It is not a perfect comparison, but there are profound similarities. Ultimately, the story of standard time offers lessons about how knowledge is constructed, but also about how knowledge is shared and accessed, and the consequences when that access is inequitably distributed. Shared information is a powerful force for change. And that is a timely reminder.

Resolutions Passed by the IMC

1　That it is the opinion of this Congress that it is desirable to adopt a single prime meridian for all nations, in place of the multiplicity of initial meridians which now exist. (Adopted unanimously.)

2　That the Conference proposes to the Governments here represented the adoption of the meridian passing through the centre of the transit instrument at the Observatory of Greenwich as the initial meridian for longitude. (Ayes, 22; noes, 1; abstaining, 2.)

3　That from this meridian longitude shall be counted in two directions up to 180 degrees, east longitude being plus and west longitude minus. (Ayes, 14; noes, 5; abstaining, 6.)

4　That the Conference proposes the adoption of a universal day for all purposes for which it may be found convenient, and which shall not interfere with the use of local or standard time where desirable. (Ayes, 23; abstaining, 2.)

5　That this universal day is to be a mean solar day; is to begin for all the world at the moment of mean midnight of the initial meridian, coinciding with the beginning of the civil day and date of that meridian; and is to be counted from zero up to twenty four hours. (Ayes, 15; noes, 2; abstaining, 7.)

6　That the Conference expresses the hope that as soon as may be practicable the astronomical and nautical days will be arranged everywhere to begin at midnight. (Adopted unanimously.)

7　That the Conference expresses the hope that the technical studies designed to regulate and extend the application of the decimal system to the division of angular space and of time shall be resumed, so as to permit the extension of this application to all cases in which it presents real advantages. (Ayes, 21; abstaining, 3.)

Notes

ABBREVIATIONS

Adams Papers	John Crouch Adams Papers, St John's College Library, Cambridge
AO	Archives of Ontario, Toronto
AGN	Archivo General de la Nacion, Santo Domingo, Dominican Republic
ARS	Archives of the Royal Society, London
BIF	Bibliothèque de l'Institut de France, Paris
BL	British Library, London
BT	BT Archives (BT = British Telecom), London
CAL	Cambridge Astronomy Library
CUL	Cambridge University Library
Fleming Papers	[Sir Sandford] Fleming Papers, MG 29, B1, Library and Archives Canada, Ottawa
IET	Institute of Engineering and Technology Archives, London
LAC	Library and Archives Canada, Ottawa
LC	Library of Congress, Washington, DC
LMA	London Metropolitan Archives
NA-UK	National Archives of the United Kingdom, London
NA-USA	National Archives of the United States of America, Washington, DC
POP	Post Office Papers, BT Archives, London
RAS	Royal Astronomical Society Archives, London
RGOA	Royal Greenwich Observatory Archives, Cambridge University Library
RGS	Royal Geographical Society Archives, London

ROE   Royal Observatory of Edinburgh
SJCL   St John's College Library, Cambridge
Smyth Archives   Charles Piazzi Smyth Archives, Royal Observatory
                       of Edinburgh
Smyth Papers   Charles Piazzi Smyth Papers, Royal Observatory of
                       Edinburgh
Strachey Papers   Papers of Lt-Gen Sir Richard Strachey, Mss. Eur F127,
                       British Library

INTRODUCTION

1 Einstein's theories of special relativity (1905) and general relativity (1916)
   completely revolutionized the way we understand time scientifically and
   philosophically. The most accessible explanation of relativity and time dilation
   can be found in Hawking, *A Brief History of Time*.
2 The original proverb, often attributed to Napoleon Bonaparte (who was
   paraphrasing), says that "history" is a series of lies agreed upon.
3 See Nanni, *The Colonisation of Time*. This trend is sometimes called the
   "Science of Inventory"; Zeller, *Inventing Canada*, 269. Donald Mackenzie
   suggests that Victorian science sought to bring a sense of order to complex
   systems; Donald Mackenzie, *Statistics in Britain*.
4 See Mitchell, *Rule of Experts*, 84–93. Mitchell argues that land surveys and
   maps of Egypt made by Europeans did not produce a more accurate view of
   the world, but rather redistributed forms of knowledge that helped colonial
   administrators exploit the natural resources of the area.
5 Glennie and Thrift, *Shaping the Day*, 40.
6 Charles Withers points out that the resolutions of the conference were not
   binding, leading "to particular and geographically uneven consequences in the
   wake of the meeting"; Withers, *Zero Degrees*, 216. Ian Bartky writes that the
   countries represented at the IMC did not unanimously agree to Greenwich as
   the prime meridian, nor did they establish time zones; Bartky, *One Time Fits
   All*, 99. Vanessa Ogle's book on time reform barely mentions the IMC – she
   makes the case that time reform was an utter failure and that the IMC in
   particular was "almost meaningless as the process of time unification dragged
   on until the 1930s and 1940s"; Ogle, *The Global Transformation of Time*, 14.
   Adam Barrows argues that the IMC did not "achieve the goal for which it is
   credited"; Barrows, *The Cosmic Time of Empire*, 36.
7 See Ogle, *The Global Transformation of Time*.
8 Astronomers had the tools and expertise required to measure sidereal time and
   convert it to solar time. The sidereal day is the length of time it takes for a star
   passing overhead to return to that spot the next night as the earth rotates. The

solar day is measured the same way, but uses the position of the sun instead of a star. Due to the earth's movement around the sun, the sidereal day is about four minutes shorter than the solar day. Once the exact time was determined, some form of time distribution was required, whether by telegraph, radio, or physically transporting accurate chronometers. All of this was expensive and complicated, making a global system of standard time much more reliant on astronomical expertise than simple local time had been.

CHAPTER ONE

1   See Blaise, *Time Lord*, 75–7. Burpee, *Sandford Fleming*, 211–12; and Lorne Green, *Chief Engineer*, 56–7. These biographies all refer to the missed train. Ian Bartky has levelled some criticism at them for this. See Bartky, *One Time Fits All*, 51, and "Sandford Fleming's First Essays on Time," 5.
2   Fleming, *Uniform Non-Local Time*, 5.
3   Bartky, *One Time Fits All*, 51, and "Sandford Fleming's First Essays on Time," 6–8. Bartky points out that the first paper was not published until 1878, two years, rather than "a few weeks," after the train incident, as Fleming claimed (Bartky suggests "a few weeks" was merely a literary device). It seems likely to me that Fleming did write the paper shortly after the train incident and simply did not publish it for several years, when, feeling pressed to publish before Cleveland Abbe and the AMS, he took steps to circulate his ideas. He did try to present a paper about time reform with the BAAS in 1876, and his own personal bibliography states that a "Memoir on Uniform, Non-Local Time" was published in 1876 (an inaccurate claim, but perhaps this is merely describing the paper he tried to present at the BAAS; it was written in 1876). See Burpee, Bibliography of Sir Sandford Fleming, 27 March 1907, vol. 7, Fleming Papers, MG 29 B1, Library and Archives Canada, Ottawa (hereinafter Fleming Papers). Fleming also told Charles Dowd in 1883 that he wrote it after the train incident in 1876, but again he may have been motivated to give the earlier date when the question of credit was in the air. Sandford Fleming to Charles Dowd, 11 Dec. 1883, vol. 65, Fleming Papers.
4   Allen, *Short History of Standard Time*.
5   Bartky, "Sandford Fleming's First Essays on Time."
6   For Struve's contribution, see Otto Struve, "First Meridian," vol. 30, Fleming Papers. Airy's time service, begun in 1849, is mentioned in H. Spencer Jones, "Untitled History of Time Service," RGOA, RGO9.625, CUL.
7   Bartky, "Sandford Fleming's First Essays on Time," 6–8.
8   Ibid., 8. Speakers included William Thomson (later Lord Kelvin), David Gill, George Darwin, the hydrographer of the Royal Navy, Robert Ball, and the Earl of Rosse.

9 Ibid., 8.
10 Abbe credited the idea of time zones to mathematician Benjamin Pierce. Bartky, "The Adoption of Standard Time," 37, note 42.
11 Ibid., 35–7. Incidentally, Bartky's article is one of the earliest publications to recognize an astronomical impetus for standard time (via Abbe's aurora measurements) alongside the needs of the railways. It does not, however, elucidate the crucial difference in scope: a specialized time for science v. a universal civil time for all people/all railway users.
12 Cleveland Abbe to Sandford Fleming, 10 March 1880 (two letters sent that date, both relevant), vol. 1, Fleming Papers; F.A.P. Barnard to Sandford Fleming, 18 March 1880, vol. 3, Fleming Papers; F.A.P. Barnard to Sandford Fleming, 23 June 1880, ibid.; F.A.P. Barnard to Sandford Fleming, 6 July 1880, ibid. See also Bartky, "Sandford Fleming's First Essays on Time."
13 Memorial of the Canadian Institute on Time Reckoning and a Prime Meridian, 1878–79. In Fleming, *Universal or Cosmic Time*, YA.2003.A.17994, BL.
14 M.E. Hicks-Beach to the Marquis of Lorne, 15 Oct. 1879, in Fleming, *Universal or Cosmic Time*. Colonial Secretary Hicks-Beach was likely paraphrasing George Airy.
15 A few decades later, this would change. The introduction of daylight saving time could be considered an experiment in enforcing changes in social behaviour. See Bartky, *One Time Fits All*, 161–200.
16 Council Minutes, 14 Nov. 1879, No. 2, vol. 8, Royal Astronomical Society Papers, Part 1, RAS.
17 Sir John Henry Lefroy's Report on Mr Sandford Fleming's Proposals Respecting a Prime Meridian and Time Reckoning, 19 Nov. 1879, RGS/CB6/1377, RGS.
18 The Secretary of the Royal Society to the Colonial Office, 6 Nov. 1879, in Fleming, *Universal or Cosmic Time*.
19 The Admiralty also rejected the proposal on the basis that the public was not yet ready for such a change. The Lords Commissioners of the Admiralty Board to the Secretary of State for the Colonies, 4 Oct. 1879, in ibid.
20 Charles Piazzi Smyth to Colonial Office, 5 Sept. 1879, in ibid.
21 Ibid.
22 Ibid.
23 Ibid.
24 George Airy to the Secretary of State for the Colonies, 18 June 1879, and Charles Piazzi Smyth to Colonial Office, 5 Sept. 1879, in ibid. In 1880, Cleveland Abbe pointed out to Fleming that the Latin alphabet is not universal, and other alphabets and writing systems were not represented in his proposal. F.A.P. Barnard to Sandford Fleming, 16 July 1880, vol. 2, Fleming Papers.

25 Withers, *Zero Degrees*, 3.
26 Smyth had seen it too. Charles Piazzi Smyth to Sandford Fleming, 12 Nov. 1878, vol. 2, Fleming Papers.
27 George Airy to Sandford Fleming, 11 Feb. 1878, vol. 1, Fleming Papers.
28 Howse, *Greenwich Time*, 114.
29 Chapman, "Sir George Airy (1801–1892)," 325.
30 Bartky, *One Time Fits All*, 70.
31 Ibid., 59.
32 About 90 per cent of the questionnaires went to "practical railroad men" and the rest to "men of a theoretical turn of mind," in other words, astronomers and academics. John Bogart to Sandford Fleming, 13 March 1882, vol. 2, Fleming Papers.
33 Vol. 2, Fleming Papers, is full of survey responses calling for some kind of reform.
34 Cleveland Abbe to Sandford Fleming, 10 March 1880, vol. 1, Fleming Papers.
35 Thomas Egleston to Sandford Fleming, 20 Feb. 1883, vol. 14, Fleming Papers.
36 Bartky, *One Time Fits All*, 59–85.
37 Mario Creet and Ian Bartky's otherwise-excellent reviews of Fleming's activism both omit his relationship with the IPAWM. Creet, "Sandford Fleming," 66–89; Barkty, *One Time Fits All*.
38 See, for example, Perry, *The Story of Standards*; Theodore Porter, *Trust in Numbers*; Wise, ed., *The Values of Precision*; Hacking, *The Taming of Chance*; Headrick, *When Information Came of Age*.
39 For more on the notion of scientific advancement facilitating colonialism, see Adas, *Machines as the Measure of Men*; Drayton, *Nature's Government*; Weaver, *The Great Land Rush*; Mitchell, *Rule of Experts*.
40 In 1884 it had about six hundred members. Reisenauer, "'The Battle of the Standards,'" 969.
41 Thomas Egleston to Sandford Fleming, 9 June 1881, vol. 14, Fleming Papers; Thomas Egleston to Sandford Fleming, 19 June 1881, ibid.; Thomas Egleston to Sandford Fleming, 24 June 1881, ibid., F.A.P. Barnard to Sandford Fleming, 11 June 1881, vol. 3, Fleming Papers.
42 F.A.P. Barnard to Sandford Fleming, 11 June 1881, vol. 3, Fleming Papers.
43 Thomas Egleston to Sandford Fleming, 1 June 1883, vol. 14, Fleming Papers.
44 Thomas Egleston to Sandford Fleming, 1 June 1883, ibid..
45 Charles Latimer to Charles Piazzi Smyth, 1 Jan 1882, A14/66, Charles Piazzi Smyth Papers, Royal Observatory of Edinburgh (hereinafter Smyth Papers). Latimer then told Fleming that although he thought the pyramid would be the best prime meridian, Greenwich might be acceptable. Charles Latimer to Sandford Fleming, 27 Feb. 1882, vol 27, Fleming Papers.
46 Charles Latimer to Sandford Fleming, 22 Dec. 1882, ibid.

47 Sandford Fleming to A.G. Wood, 14 Feb. 1883, vol. 54, Fleming Papers; Sandford Fleming, "Standard Time," *International Standard* 1, undated, vol. 65, Fleming Papers; *International Standard*, March 1883, vol. 105, Fleming Papers.

48 Charles Latimer to Sandford Fleming, 30 Oct. 1883, vol. 27, Fleming Papers; Charles Latimer to Sandford Fleming, 22 Nov. 1883, ibid.; Charles Latimer to Sandford Fleming, 5 Dec. 1883, ibid.

49 F.A.P. Barnard to Sandford Fleming, 4 June 1881, vol. 3, Fleming Papers, See also Bartky, *Selling the True Time*, 149.

50 Bartky, *One Time Fits All*, 35–47.

51 Charles Piazzi Smyth to F.A.P. Barnard, 24 Aug. 1881, vol. 3, Fleming Papers.

52 George Airy to F.A.P. Barnard, 12 July 1881, ibid.

53 F.A.P. Barnard to Sandford Fleming, 30 July 1881, ibid. See also F.A.P. Barnard to Sandford Fleming, 3 Sept. 1881, ibid.

54 F.A.P. Barnard to Sandford Fleming, 19 Dec. 1881, ibid.

55 His remarks to the Congress at Venice can be found in Fleming, *The Adoption of a Prime Meridian*.

56 George Wheeler to Sandford Fleming, 2 March 1882, vol. 53, Fleming Papers.

57 Ibid. .

58 Sandford Fleming to John Bogart, 26 Oct. 1881, vol. 63, Fleming Papers. See also Bartky, *One Time Fits All*, 66–7.

59 Sandford Fleming to Charles Tupper, 20 Oct. 1883, vol. 65, Fleming Papers.

60 "The Geodetic Conference at Rome," *Journal of the Society of Arts* 32, no. 1625 (Friday 11 Jan. 1884): 132–3.

61 Report by the Committee, undated, vol. 188, Papers of Lt-Gen Sir Richard Strachey, Mss. Eur F127, British Library (hereinafter Strachey Papers).

62 The Treasury and the warden of the standards, H.J. Chaney, were particularly opposed.

63 Bartky, "Inventing," 111.

64 If Europeans and North Americans understood timekeeping differently, geography is probably somewhat to blame. North America's enormous width made standardizing time more pressing there. Fleming wrote in an early paper that people in Europe could not understand the problem faced in the United States or Russia, because of the differences in longitude. Sandford Fleming, *Time Reckoning*, 1879, Canadian and U.S. Papers on Time Reckoning, E.13.1, CAL.

65 Born in Nova Scotia, Newcomb moved about 1854 to the United States, where he became a professor of mathematics and astronomy at Johns Hokpins University, served in the U.S. Naval Observatory, and oversaw the Nautical Almanac Office.

66 See Pietsch, *Empire of Scholars*; Edney, *Mapping an Empire*.

67 See Time Service, vol. 14523, RG 30, LAC. The whole file is all about the
    "day of two noons."
68 See O'Malley, *Keeping Watch*, 118–19, 126, 130–44; Howse, *Greenwich Time*,
    126.
69 *Indianapolis Sentinel*, 21 Nov. 1883.
70 A Toronto Woman to Sandford Fleming, 19 Nov. 1883, vol. 54, Fleming
    Papers.
71 Howse, *Greenwich Time*, 114.
72 See G.W. Wicksteed to Sandford Fleming, 9 July 1891, vol. 53, Fleming Papers;
    G.W. Wicksteed to Sandford Fleming, 12 Jan. 1892, ibid. The U.S. government
    didn't standardize time all over the nation until 1918; Howse, *Greenwich
    Time*, 126. In Canada, it remained up to the provinces; Thomson, *The
    Beginning of the Long Dash*, 34.
73 G. Powell to Sandford Fleming, 8 May 1883, Canadian Institute Papers, file
    4-0-2, F1052, AO.
74 Sandford Fleming to Charles Tupper, 9 May 1883, vol. 65, Fleming Papers.
75 Sandford Fleming to G. Powell, 23 June 1884, ibid.
76 Foreign Office Letterbooks, Entry for 28 Nov. 1882, Foreign Office Papers,
    FO566.19, NA-UK.
77 Foreign Office to Colonial Office, 5 June 1883, Colonial Office Papers,
    CO42.776, NA-UK.
78 Lionel West to Frederick Frelinghuysen, 14 June 1883, vol. 27, Fleming Papers;
    Lionel West to Frederick Frelinghuysen, 8 June 1883, ibid.; Frederick
    Frelinghuysen to Lionel West, 13 June 1883, ibid.
79 Adoption of a Common Prime Meridian, 6 June 1883, Colonial Office Papers,
    CO42.776, NA-UK; Colonial Office to Foreign Office Draft, 7 June 1883, ibid.
80 Sandford Fleming to Charles Tupper, 14 July 1883, Colonial Office Papers,
    CO42.775, NA-UK.
81 The Earl Granville to Mr Lowell, 21 July 1883, vol. 11, Fleming Papers.
82 This military intervention, in which the Canadian government quelled a
    Métis rebellion led by Louis Riel, has a controversial legacy. As an act of col-
    onial violence, it had ill consequences for the Métis people in the (then)
    NorthWest Territories and alienated much of the francophone population in
    Canada. In English-speaking Canada at the time, however, the fight was well
    received, and use of the railway to deploy troops quickly made the Canadian
    Pacific Railway very popular, allowing the government to finish the line. Both
    Fleming and Tupper approved of the use of force. In Tupper's words, it was
    necessary in "defence of a country that you and I have laboured so earnestly
    to build up." Charles Tupper to Sandford Fleming, 25 June 1885, vol. 50,
    Fleming Papers. .

83 Adoption of the Multiple of Greenwich Time in Canada and the United States, 15 Nov.1883, Colonial Office Papers, CO42.775, NA-UK.

84 Sandford Fleming to Charles Tupper, 20 Oct. 1883, ibid.

85 Science and Art Department to Colonial Office, 16 Jan. 1884, vol. 11, Fleming Papers.

86 Treasury to Foreign Office, 14 Jan. 1884, Foreign Office Papers, FO5.1886, NA-UK.

87 Foreign Office to Treasury, 21 Jan. 1884, ibid.

88 Colonial Office to Treasury, 7 Feb. 1884, ibid.

89 John Donnelly to William Christie, 16 Feb. 1884, RGOA, RGO7.142, CUL. Donnelly was a military officer, but spent most of his later career reforming the SAD and helped create scientific-education programs. See Vetch, "Donnelly."

90 Royal Society to Foreign Office, 7 Feb. 1884, Foreign Office Papers, FO5.1886, NA-UK.

91 Council Minutes, 7 Feb. 1884, Royal Society Council Minutes, CMO17, ARS.

92 Treasury to Foreign Office, 13 Feb. 1884, Foreign Office Papers, FO5.1886, NA-UK.

93 Science and Art Department to Foreign Office, 26 Feb. 1884, ibid.

94 Memo Science and Art Department, 26 Feb. 1884, ibid.; Cecil Spring-Rice to Sanderson, 4 March 1884, ibid.

95 Cecil Spring-Rice to Sanderson, 13 Feb. 1884, ibid. The SAD's offices were located in South Kensington, in "Albertopolis," the vast educational and cultural complex that flowed from the profits of the Great Exhibition of 1851, held just north in the Crystal Palace in Hyde Park and headed by Prince Albert.

96 Initialled Memo, 28 Feb. 1884, ibid.

97 Treasury to Foreign Office, 7 March 1884, ibid. The amount decided on was £100 per delegate.

98 Science and Art Department to Foreign Office, 3 May 1884, ibid.

99 Colonial Office to Foreign Office, 2 May 1884, ibid.

100 Science and Art Department to Foreign Office, 10 May 1884, ibid.

101 Lowell to Earl Granville, 21 May 1884, Foreign Office Papers, FO5.1884, NA-UK.

102 Prime Meridian Conference, Foreign Office Minutes, 26 May 1884, Colonial Office Papers, CO42.779, NA-UK.

103 Prime Meridian Conference at Washington, Foreign Office Minutes, 18 April 1884, ibid.

104 Ibid. In 1883, Australian newspapers caught wind of German plans to annex New Guinea. In Queensland, there was a public outcry for Britain to annex it instead. Colonial Secretary Lord Derby shot down the idea, to the Australians' displeasure.

105 Science and Art Department to Foreign Office, 30 May 1884, Foreign Office
     Papers, FO5.1886, NA-UK.
106 Science and Art Department to Foreign Office, 21 June 1884, Foreign Office
     Papers, FO5.1887, NA-UK.
107 Colonial Office to Foreign Office, 24 June 1884, ibid.
108 Foreign Office to Colonial Office Draft, 28 June 1884, ibid.
109 Prime Meridian Conference, Colonial Office Minutes, 28 June 1884, Colonial
     Office Papers, CO42.779, NA-UK.
110 Ibid.
111 Ibid.
112 Ibid.
113 Ibid. The Treasury of course knew about the proposed conference, directly
     from the U.S. government, but had decided not to send delegates. Fleming's
     request may have been the first time the CO had heard of it.
114 Ibid.
115 Sandford Fleming to G. Powell, 23 June 1884, vol. 65, Fleming Papers;
     Sandford Fleming to Charles Tupper, 23 June 1884, ibid.
116 G. Powell to Sandford Fleming, 18 July 1884, vol. 39, Fleming Papers.
117 Charles Tupper to Sandford Fleming, 17 July 1884, vol. 50, Fleming Papers.
     The word "projector" is nearly illegible, and may also be "proprietor,"
     "progenitor," and so on. The word is "originator" in Colonial Office
     Letterbooks, Miscellaneous Correspondence, 15 July 1884, Prime Meridian
     Conference, Colonial Office Papers, CO340.2, NA-UK.
118 Governor of South Australia to Colonial Office, 9 July 1884, Colonial Office
     Papers, CO201.601, NA-UK; Colonial Office Letterbooks, Miscellaneous
     Correspondence, 28 Sept. 1884, Prime Meridian Conference, Colonial Office
     Papers, CO340.2, NA-UK.
119 Prime Meridian Conference, Colonial Office Minutes, 6 Oct. 1884, Colonial
     Office Papers, CO201.601, NA-UK. The CO had asked about the conference
     through South Australia's government, instead of directly asking each colony
     separately.
120 Science and Art Department to Colonial Office, 22 July 1884, Foreign Office
     Papers, FO5.1887, NA-UK.
121 Ibid.
122 Ibid.
123 Lowell to Frederick Frelinghuysen, 29 Aug. 1884, Despatches from U.S.
     Ministers to Great Britain 1791-1906, Microfilm M30 146, NA-USA.
124 Ibid.; U.S. Department of State to F.A.P. Barnard, 18 Sept. 1884, Domestic
     Letters of the Department of State 1784–1906, Microfilm M40 101, NA-USA.
125 Lowell to Colonial Office, 5 Sept. 1884, Colonial Office Papers, CO42.779,
     NA-UK.

126 Prime Meridian Conference, Colonial Office Minutes, 16 Sept. 1884, ibid.
127 Colonial Office to Science and Art Department Draft, 16 Sept., ibid.
128 Prime Meridian Conference, Colonial Office Minutes, 29 Aug. 1884, ibid.
129 Prime Meridian Conference, Colonial Office Minutes, 20 Aug. 1884, ibid. In this case the SAD had thought that the CO was asking it to take away the vote from Adams, Evans, or Strachey and give it to Fleming. The CO was simply trying to make sure that one of the two new delegate positions now allowed would be saved for Fleming.
130 Hirsch was slated to represent Switzerland at the IMC, but it is unclear whether he was able to attend. If he did attend, he did not speak at the conference.
131 Ad.[olphe] Hirsch to C.W. Siemens, 13 Jan. 1883, vol. 188, Strachey Papers.
132 Wilhelm Foerster to William Christie, 4 May 1884, RGOA, RGO7.147, CUL.
133 David Gill to William Christie, 25 April 1884, RGOA, RGO7.148, CUL.
134 William Christie to Peter MacLiver, 1 May 1884, ibid.
135 William Christie to John Donnelly, 19 Feb. 1884, RGOA, RGO7.142, CUL; see also William Christie to Peter MacLiver, 1 May 1884, RGOA, RGO7.148, CUL.
136 William Christie to Peter MacLiver, 10 May 1884, RGOA, RGO7.148, CUL; William Christie to William Foerster, 20 May 1884, ibid.
137 Treasury to the President of the Royal Society, 23 May 1884, ibid.
138 Ibid.
139 William Christie to John Donnelly, 22 Feb. 1884, RGOA, RGO7.142, CUL.
140 William Christie to John Donnelly, 28 March 1884, ibid.
141 Association Geodésique Internationale to William Christie, 1 April 1884, RGOA, RGO7.148, CUL. In the late 1870s, British-French relations concerning Egypt had been reasonably amicable. The country lay geographically between their imperial interests (France in northern Africa and Britain via the Suez in India), and a neutral Egypt kept the peace between the two powers. Unrest in Egypt in 1881–82 persuaded French and British officials to send in warships in an attempt to deter the uprising. When that failed, the British sent in troops, which was a step beyond what France was willing to do. The fervently imperialist French president, Jules Ferry, had just been ousted, and his country's politics was in turmoil. Ferry was re-elected in early 1883, too late to involve France in the Egyptian conflict, but his expansionist stance fostered resentment towards Britain's unilateralism there. The result was an Egypt effectively annexed by Britain, and a France disillusioned with its neighbour's friendship. Britain struggled to maintain order in Egypt over the next several years. The crisis in Sudan in 1883–84 made its rule in Egypt seem particularly incompetent, fuelling international criticism of its intervention. All of this, in the larger

context of the renewed scramble for Africa, made French-British relations highly volatile by 1884. See Stone and Otte, *Anglo-French Relations*; Mansfield, *The British in Egypt*.

142 David Gill to William Christie, 25 April 1884, RGOA, RGO7.148, CUL.

143 William Christie to Ad.[olphe] Hirsch, 12 April 1884, ibid.

144 William Christie to Richard Strachey, 18 July 1884, vol. 187, Strachey Papers.

145 William Christie to Ad.[olphe] Hirsch, 16 Sept. 1884, RGOA, RGO7.148, CUL.

146 John Donnelly to Richard Strachey, 24 June 1884, vol. 188, Strachey Papers.

147 H.F. Anson to Sandford Fleming, 25 Sept. 1884, vol. 27, Fleming Papers. Fleming was concerned about his letter of appointment arriving late. A friend wrote to him on 26 September: "I hope therefore that your papers will be in time." A. McLellan to Sandford Fleming, 26 Sept. 1884, vol. 33, Fleming Papers.

148 Prime Meridian Conference, Colonial Office Minutes, 1 Sept. 1884, RGOA, CO42.779, CUL.

149 Prime Meridian Conference, Colonial Office Minutes, 3 Oct. 1884, RGOA, CO201.601, ibid.

150 Prime Meridian Conference, Colonial Office Minutes, 28 Sept. 1884, RGOA, CO309.127, ibid. Cockle would not have been able to arrive in Washington until 30 October, at the earliest.

151 Earl Granville to J.R. Lowell, 2 Oct. 1884, Despatches from U.S. Ministers to Great Britain 17911906, Microfilm M30 146, NA-USA.

152 Bartky, *One Time Fits All*, 83.

153 An Open Letter to the President of the United States from the IPAWM, 30 July 1884, Letters to the International Meridian Conference of 1884, box 1, Records of International Conferences, Commissions, and Expositions, RG 43, NA-USA.

154 Charles Latimer to Sandford Fleming, 21 Aug. 1884, vol. 27, Fleming Papers.

155 Charles Latimer to Sandford Fleming, 29 Sept. 1884, ibid.

156 John Bogart to Sandford Fleming, 27 Aug. 1884, vol. 2, Fleming Papers.

157 John Bogart to Sandford Fleming, 20 Sept. 1884, ibid.

158 F.A.P. Barnard to Sandford Fleming, 21 Sept. 1884, vol. 3, Fleming Papers.

159 Jules Janssen to Henriette Janssen, 27 Sept. 1884 (my translation), Correspondance de Jules Janssen, Ms. 4133, BIF.

160 Jules Janssen to Henriette Janssen, 27 Sept. 1884, ibid.

161 Rutherfurd was trained as a lawyer but retired early to devote his time to his passion, astronomy. He devoted considerable effort to make a precision screw for a spectroscopic instrument for physical studies of stars. Lewis Rutherfurd to Simon Newcomb, 24 July 1884, box 38, Newcomb Papers, LC.

CHAPTER TWO

1 For more on the transit-of-Venus preparations, see Ratcliff, *The Transit of Venus Enterprise*; Forbes, *The Transit of Venus*, 63–7.
2 Forbes, *The Transit of Venus*.
3 Heathorn, *For Home*, 166.
4 Gillard, "Education in England."
5 Earlier in the nineteenth century, girls were more likely to be taught French than boys. See Tomalin, *The French Language*, 79.
6 Withers, "Scale," 99.
7 Cooper, *Colonialism in Question*, 91. See also Potter, "Webs."
8 Pietsch, *Empire of Scholars*.
9 See Visram, *Asians in Britain*; Burton, *At the Heart of the Empire*.
10 Fisher, *Counterflows to Colonialism*, 299.
11 See Tabili, *We Ask for British Justice*; Tabili, "A Homogeneous Society?"
12 Butler, *Gender Trouble*, 173.
13 Shively, *Tradition and Modernization*; Irokawa, *The Culture of the Meiji Period*.
14 See Davidoff and Hall, *Family Fortunes*.
15 See ibid.; Vickery, "Golden Age"; Anna Clark, *The Struggle*; Levine-Clark, *Beyond the Reproductive Body*. Levine-Clark suggests that working-class women did not share this ideal.
16 Watts, *Women in Science*, 103.
17 See Butler, *Gender Trouble*, 173. "The gendered body is performative."
18 Watts, *Women in Science*, 124.
19 Walter Maunder, *The Astronomy of the Bible*, 271–2.
20 Draper, *History of the Conflict*.
21 Garwood, *Flat Earth*, 11–12.
22 Ibid, 60–1.
23 See Mary Brück, "Lady Computers at Greenwich," 86–7. Russell was hired to replace Furniss, who left after a year. Rix left soon after for health reasons. Clemes was older than the others, and was initially considered their supervisor. See Dolan, "Christie's Lady Computers."
24 These women were crassly known as "Pickering's Harem" (after the observatory's director). Over eighty had worked with him by 1918. As they were paid less than men, he could hire more of them. See Grier, *When Computers Were Human*, 82–4.
25 Lady Computers, RGOA, RGO7.140, CUL.
26 Ibid. For more on boy computers, see Johnston, "Managing the Observatory," 155–75; Schaffer, "Astronomers Mark Time," 115–45; and Aubin, "On the Epistemic and Social Foundations."

27 H.H. Turner to Fanny Allen, 18 Feb. 1892, RGOA, RGO7.140, CUL.

28 Russell's letter of recommendation upon leaving lists various duties and responsibilities beyond computing, including using the observatory's instruments. William Christie re. Annie Russell, c. 1897, RGOA, RGO7.138, CUL.

29 At least one woman who applied turned down the job because of its low salary. L.S. Walter to H.H. Turner, 22 Dec. 1891, RGOA, RGO7.140, CUL.

30 Annie Russell to H.H. Turner, c. 1891, ibid. She was offered £48 a year, while the school had paid £80 plus residence. Her salary eventually increased slightly.

31 Lady Computers.

32 Annie (Russell) Maunder to Dr Dyson, 4 Dec. 1914, RGOA, RGO8.150, CUL.

33 Mary Brück, "Lady Computers," note 42. "Women in the public service were required to resign upon marriage."

34 Annie Russell to William Christie, Sept. 1895, RGOA, RGO7.138, CUL. Annie Russell changed her name to Annie Maunder after marriage. For ease of understanding, I continue to refer to her as Russell to distinguish from her husband, Walter Maunder.

35 See Jalland, *Women, Marriage, and Politics*, 189, 195–204. Jalland discusses women as political confidantes. For more on couples in science, see Pycior, Slack, and Abir-Am, eds., *Creative Couples in the Sciences*; and Lykknes, Opitz, and Tiggelen, eds., *For Better or Worse*.

36 Lettres écrites, au cours de ses nombreux voyages, par Jules Janssen à sa femme Henriette, Correspondance de Jules Janssen, Ms. 4133, 273–80, BIF.

37 Ogilvie, "Obligatory Amateurs," 83.

38 Russell was the lead author of *The Heavens and Their Story* and a significant contributor to *The Astronomy of the Bible*. Maunder wrote in the dedication of the latter, "To my wife, my helper in this book and in all things."

39 Maunder and Maunder, *The Heavens and Their Story*, 26, 35.

40 Ibid., 8, 26.

41 Ibid. 228–9.

42 Ibid., 9.

43 Walter Maunder, *Astronomy of the Bible*, 269.

44 Ibid., 271.

45 Ibid., 273–4.

46 Ibid., 400.

47 See Laughton, "Snow."

48 Snow, "Ocean Relief Depots," 753–5.

49 Ibid.

50 Snow, "An International Prime Meridian," Circular Letter, Mic.A.19863, BL.

51 Ibid.

52 The St Peter and St Paul Archipelago.

53 Snow, "An International Prime Meridian."

54 Ibid.

55 Tosh, *Manliness and Masculinities*, 42.

56 Pionke, *The Ritual Culture*.

57 Levine, *The Amateur and the Professional*, 124.

58 Lankford, "Amateurs versus Professionals," 12. See also Meadows, "Lockyer"; Endersby, *Imperial Nature*, 1–2.

59 Lorimer, *Science*, 114.

60 Lankford, "Amateurs versus Professionals," 12.

61 They cost £7 10 s – almost two months' wages for Annie Russell's computer job at Greenwich. That amount was approximately the monthly wage of a highly skilled working man in the 1880s; most skilled workers earned £6–£8 a month; unskilled workers £3–£5.

62 Josiah Latimer Clark, *Transit Tables*.

63 Ibid.

64 Lankford, "Amateurs versus Professionals," 28.

65 Pionke, *The Ritual Culture*.

66 Charles Piazzi Smyth, Journal, 24 Nov. 1884, Journal 36, Charles Piazzi Smyth Archives, Royal Observatory of Edinburgh (hereinafter Smyth Archives).

67 There are plenty of requests for advice in Correspondence A12/52, A12/54, A13/58, Smyth Papers; George Airy forwarded questions about timekeeping from the German ambassador to Smyth for answers in 1878, RGOA, RGO6.13, CUL. See also Brück and Brück, *The Peripatetic Astronomer*, 256–7.

68 See Reisenauer, "'The Battle of the Standards,'" 943–4.

69 Smyth, *Our Inheritance*, 39.

70 Ibid., 39.

71 Ibid., 339.

72 Ibid., 115.

73 Ibid., 291.

74 Ibid, 338.

75 Ibid., 351.

76 Ibid., 552.

77 Reisenauer, "'The Battle of the Standards,'" 950.

78 Gange, *Dialogues with the Dead*, 1-2.

79 Perhaps the most sought-after prizes were mummies. The voyeuristic spectacle of mummy "unwrappings" became popular entertainment. Public shows, sometimes accompanied by a brass band, drew audiences from the working classes, while private parties with professional surgeons and dissectors entranced the wealthy. Parramore, *Reading the Sphinx*, 27, 30–1; France, *The Rape of Egypt*, 174.

80 Parramore, *Reading the Sphinx*, 34–5.
81 Gange, *Dialogues with the Dead*, 2. The Pithom stele was believed to relate to the biblical Exodus of the Israelites from Egypt.
82 Ibid., 154.
83 See Said, *Orientalism*.
84 Waynman Dixon to Charles Piazzi Smyth, 7 Jan. 1877, A13/58, Smyth Papers.
85 Smyth, *Our Inheritance*, ix.
86 Gange, *Dialogues with the Dead*, 152–3.
87 Anderson, "The Development of British Tourism in Egypt."
88 Charles Piazzi Smyth, Report of the Royal Observatory of Edinburgh, 30 June 1888, A17/95, Smyth Papers.
89 Jon Smythe to Charles Piazzi Smyth, 24 May [1877?], A13/59, ibid.
90 A. Bedford to Charles Piazzi Smyth, 12 Aug. 1882, A14/66, ibid.
91 Nicola Mary Belham to Jessica Smyth, 1 May 1877, A13/59, ibid.
92 Charles Piazzi Smyth copied a poem by John Stuart Blackie about Gordon from *The Leisure Hour* into his journal, July 1885, Journal 37, Smyth Archives.
93 J.C. Adams to Rev. Bashforth, 21 Nov. 1884, 4/26/3–4, John Couch Adams Papers, St John's College Library, Cambridge (hereinafter Adams Papers).
94 Waynman Dixon to Charles Piazzi Smyth, 7 Jan. 1877, A13/58, Smyth Papers.
95 Waynman Dixon to Charles Piazzi Smyth, 7 Jan. 1877, A13/58, ibid.
96 Reisenauer, "'The Battle of the Standards,'" 955.
97 See Correspondence, A12/52, A13/59, Smyth Papers.
98 Reisenauer, "'The Battle of the Standards,'" 956.
99 Charles Latimer to Charles Piazzi Smyth, 20 Feb. 1880, A14/65, Smyth Papers.
100 Reisenauer, "'The Battle of the Standards,'" 37.
101 Charles Latimer to Charles Piazzi Smyth, 8 March 1880, A14/65, Smyth Papers.
102 See Brück and Brück, *The Peripatetic Astronomer*, 133; Reisenauer, "'The Battle of the Standards,'" 28–33.
103 Reisenauer, "'The Battle of the Standards,'" 962.
104 The notion that imperial sentiment reached new heights in Britain from the 1880s on is widely accepted, although its exact nature is hotly debated. See John Mackenzie, *Propaganda and Empire*; Bernard Porter, *The Absent-Minded Imperialists*; Andrew Thompson, *The Empire Strikes Back*.
105 Reisenauer, "'The Battle of the Standards,'" 948.
106 See Brück and Brück, *The Peripatetic Astronomer*, 119; Reisenauer, "'The Battle of the Standards,'" 953.
107 Reisenauer, "'The Battle of the Standards,'" 972–3.
108 Brück and Brück, *The Peripatetic Astronomer*, 229; Petrie, *The Pyramids and Temples of Gizeh*, 189.

109 Smyth himself never espoused British Israelism, but he engaged with that community because it supported his theories. See Reisenauer, "'The Battle of the Standards,'" 956. Smyth attempted to convince the War Office to remeasure the pyramid after Egypt was invaded in 1882, to no avail. See Charles Piazzi Smyth Journal, 24–27 Feb. 1882, Journal 34, Smyth Archives.

110 "The Great Pyramid," *Daily Review* (1869), RGOA, RGO6/365, CUL.

111 Quoted in Reisenauer, "'The Battle of the Standards,'" 957. See also Charles Piazzi Smyth's Letter of Resignation, 7 Feb. 1874, A12/55, Smyth Papers.

112 Charles Piazzi Smyth's Letter of Resignation, 7 Feb. 1874, A12/55, Smyth Papers. Smyth may have meant his resignation only as a threat. A decade earlier, he had made a similar threat to the Royal Society of Edinburgh, when one of his early critics, James Simpson, published disparaging remarks about Smyth's pyramid theories in its *Proceedings*, but the president visited Smyth and talked him out of it. This time, however, no one tried to stop him. Instead, he was forced to leave, now thoroughly convinced that the Royal Society had mistreated him. See Brück and Brück, *The Peripatetic Astronomer*, 123, 177–8; "The Great Pyramid," *Daily Review* (1869), RGOA, RGO6/365, CUL.

113 Reisenauer, "'The Battle of the Standards,'" 957.

114 Brück and Brück, *The Peripatetic Astronomer*, 180.

115 James Napier to John Couch Adams, 23 Jan. 1876, 11/26/1, Adams Papers.

116 Charles Piazzi Smyth, Equatorial Book, vol. 2, 1879–1888, A17/95, Smyth Papers.

CHAPTER THREE

1 Constance Green, *Washington*, 363.

2 Ogle, *The Global Transformation of Time*, 35.

3 See Howse, *Greenwich Time*; Galison, *Einstein's Clocks*; Bartky, *One Time Fits All*.

4 Barrows, *The Cosmic Time of Empire*, 34.

5 E. Strachey to Richard Strachey, 6 Sept. 1884, vol. 114, Strachey Papers. To be fair to Mr West, I had to look up "geodetic" too.

6 Some of the breakdown by occupation is imperfect, as several delegates fit into more than one category: for example, some naval representatives were also astronomers, such as S.R. Franklin.

7 *Protocols of the Proceedings*, 15. These invited guests were Asaph Hall, Julius Hilgard, Simon Newcomb, William Thompson (Lord Kelvin), and Karl Wilhelm Valentiner. Hilgard was more of an engineer and surveyor.

8 Barrows, *The Cosmic Time of Empire*, 23–9; Nanni, *The Colonisation of Time*; Ogle, *The Global Transformation of Time*.

9 Barrows, *The Cosmic Time of Empire*, 23–9.

10 Nanni, *The Colonisation of Time*, 54.

11 Ibid., 2.

12 Ogle, *The Global Transformation of Time*, 204.

13 Director of the Astronomical Observatory of Bogota to the Director of the Naval Observatory of Washington, 16 Feb. 1883, Letters Received, PC42, entry 7, box 49A, Records of the United States Naval Observatory, RG 78, NA-USA.

14 Vice-Admiral, U.S. Navy, to Envoy Ex. and Min. Plen. of Colombia, Draft, 13 Oct.1884, Letters Received, ibid.

15 S.R. Franklin to Ricardo Becerra, 4 Nov. 1884, Miscellaneous Letters Sent, PC42, entry 4, vol. 5, ibid.

16 The BAAS included scientists across technical disciplines, so was a good forum for Fleming's ideas, despite his early failures with the organization.

17 Sandford Fleming, diary entry, 27 Aug. 1884, vol. 81, Fleming Papers; J.C. Adams to Francis Bashforth, 25 April 1884, 4/26/3–4, Adams Papers; Ralph Strachey to Richard Strachey, 20 July 1884, vol. 132, Strachey Papers; Henry Strachey to Richard Strachey, 29 July 1884, vol. 122, Strachey Papers. Although I have been unable to confirm whether Strachey attended the BAAS in Montreal in September 1884, he probably did. Another IMC delegate, Sir Frederick Evans, might have as well.

18 J.C. Adams to George ---, 28 Sept. 1884, 16/1/1, Adams Papers.

19 Ibid.

20 J.C. Adams to Francis Bashforth, 25 April 1884, 4/26/3–4, Adams Papers.

21 Sandford Fleming, diary entry, 30 Sept. 1884, vol. 81, Fleming Papers.

22 Sandford Fleming, diary entry, 18 May 1882, ibid.

23 Jules Janssen to Henrietta Janssen, 27 Sept. 1884, Lettres écrites, au cours de ses nombreux voyages, par Jules Janssen à sa femme Henriette, Correspondance de Jules Janssen, Ms. 4133, 273–80, BIF.

24 Manuel de Jesús Galvan to Señor Ministro de Relaciones Exteriores, 20 Sept., 13 Oct., 28 Oct., 30 Oct. 1884, vol. LIX, Textos Reunidos 4: Cartas, Ministerios y Misiones Diplomaticas, AGN.

25 Manuel de Jesús Galvan to Señor Ministro de Relaciones Exteriores, 28 Oct. 1884, vol. LIX, ibid.; Sackville West Reporting on the U.S. Presidential Election, October–November 1884, Foreign Office Papers, FO5.1872, NA-UK.

26 Valrose, *Hon. Uncle Sam*, 44.

27 Ibid., 44.

28 Keim, *Society in Washington*, 55.

29 Simon Newcomb to Otto Struve, [Feb.?] 1885, box 6, Newcomb Papers, LC.

30 Valrose, *Hon.Uncle Sam*, 45.

31 Thoron, ed., *First of Hearts*, 20.

32 "Some Foreign Ministers at Washington: Eight Portraits," 61–8.

33 Sandford Fleming, diary entry, 6–8 Oct. 1884, vol. 81, Fleming Papers.

34 Sandford Fleming, diary entry, 15 Oct. 1884, ibid.

35 *New York Times,* 17 Oct. 1884; J.C. Adams, diary entry, 16 Oct. 1884, 39/11/4, Adams Papers.

36 Invitation to Mount Vernon from Secretary of State to IMC Delegates, 16 Oct. 1884, vol. 188, Strachey Papers.

37 John Donnelly to Richard Strachey, 24 June 1884, ibid.

38 J.C. Adams to Simon Newcomb, 14 July 1884, 37/21/4, Adams Papers.

39 F.A.P. Barnard to Sandford Fleming, 21 Sept. 1884, vol. 3, Fleming Papers.

40 Ibid.

41 F.A.P. Barnard to Sandford Fleming, 2 Oct. 1884, vol. 3, Fleming Papers.

42 Ibid.

43 *Protocols of the Proceedings,* 18.

44 Ibid., 21, 74. The rest of the committee included the delegates of Germany, Mr Hinckeldeyn; of the United States, Professor Abbe; of Japan, Mr Kikuchi; and of Costa Rica, Mr Echeverria.

45 D.J. Byrne to Admiral Rodgers, undated, Letters to the International Meridian Conference 1884, box 1, Records of International Conferences, Commissions, and Expositions, RG 43, NA-USA.

46 *Protocols of the Proceedings,* 24.

47 Sandford Fleming, diary entry, 3 Oct. 1884, vol. 81, Fleming Papers.

48 Sanford Fleming, *On Uniform Standard Time, For Railways, Telegraphs and Civil Purposes Generally,* vol. 122, Fleming Papers. Fleming summarized his recommendations to the conference delegates in Fleming, *The International Prime Meridian Conference: Recommendations Suggested.*

49 *Protocols of the Proceedings,* 36.

50 Ibid., 36–7.

51 Galison, *Einstein's Clocks,* 128: Galison describes an "explosive atmosphere surrounding the Prime Meridian showdown" because of French and British competition to "possess" mapmaking. Withers describes the IMC as a conflict between British common sense (and self-interest) and French principles of scientific perfectibility and neutrality; Withers, *Zero Degrees,* 185–215. See also Howse, *Greenwich Time,* 138–51 (a perfunctory account). Bartky focuses on the American delegation; Bartky, *One Time Fits All,* 82–95.

52 *Protocols of the Proceedings,* 36.

53 *New York Times,* 8 Oct. 1884.

54 Ibid.

55 *Daily News* (London), 7 Oct. 1884.

56 C.H. Mastin to Richard Strachey, 11 Oct. 1884, vol. 151, Strachey Papers.

57 Sandford Fleming, diary entry, 4 Oct. 1884, vol. 81, Fleming Papers.

58 Sandford Fleming, diary entry, 22 Oct. 1884, ibid.

59 Jules Janssen to Henrietta Janssen, 13 Oct. 1884, Lettres écrites, au cours de ses nombreux voyages, par Jules Janssen à sa femme Henriette, Correspondance de Jules Janssen, Ms. 4133, 273–80, BIF; translation by Dunlop in Launay, *The Astronomer Jules Janssen,* 132–3.

60 Sandford Fleming, diary entry, 5 Oct. 1884, vol. 81, Fleming Papers.

61 Sandford Fleming, diary entry, 12 Oct. 1884, ibid.

62 Videira, "Luiz Cruls e o Premio Valz de Astronomia," 85–104.

63 Ogle, *The Global Transformation of Time,* 86–7.

64 *Protocols of the Proceedings,* 81.

65 Galison, *Einstein's Clocks,* 156.

66 *Protocols of the Proceedings,* 87 (emphasis added).

67 Ibid.

68 Ibid., 88.

69 Ibid., 92.

70 "Greenwich Time All Over the World," 66.

71 Manuel de Jesús Galvan to Señor Ministro de Relaciones Exteriores, 27 Oct. 1884, vol. 59, Textos Reunidos 4: Cartas, Ministerios y Misiones Diplomaticas, AGN.

72 *Protocols of the Proceedings,* 101.

73 Ibid., 116.

74 Ibid., 117–18.

75 Ibid., 121.

76 Ibid., 127.

77 Ibid., 128.

78 Ibid., 129.

79 Ibid., 134.

80 Ibid., 135.

81 Ibid., 136.

82 Ibid., 136.

83 Ibid., 146. Italics show the new words that were added.

84 Deringil, *Conversion and Apostasy,* 175.

85 Gawrych, *The Young Atatürk,* 86.

86 Ogle, *The Global Transformation of Time,* 132–3.

87 Ibid., 121, 129, 135–6.

88 For more on the Ottoman organization of time, see Wishnitzer, *Reading Clocks, Alla Turca.*

89 *Protocols of the Proceedings,* 114.

90 Ibid., 178.

91 Ibid., 179–80.

92 Ibid., 179.

93 When a later resolution decreed that the universal day would be counted from

0 to 24 hours, however, Rustem voted against it (he voted for it at first, but later voted it down). Ibid., 205.

94 Ibid., 164–5.

95 Ibid., 165.

96 Ibid., 176.

97 Ibid., 171.

98 Ibid., 182.

99 Sandford Fleming, diary entry, 21 Oct. 1884, vol. 81, Fleming Papers.

100 *Protocols of the Proceedings*, 198.

101 Sandford Fleming, diary entry, 21 Oct. 1884, vol. 81, Fleming Papers.

102 Sandford Fleming, diary entry, 22 Oct. 1884, ibid.

103 Sandford Fleming, diary entry, 23 Oct. 1884, ibid.

104 Admiral Rodgers to Sandford Fleming, 31 Oct. 1884, vol. 41, Fleming Papers.

105 *Protocols of the Proceedings*, 167.

106 J.C. Adams to Francis Bashforth, 25 April 1885, 4/26/3–4, Adams Papers.

107 J.C. Adams, diary entry, 25 Oct. 1884, 39/11/4, ibid.

108 Frederick Frelinghuysen to Manuel de Jesús Galvan, 26 Oct. 1884, vol, 59, Textos Reunidos 4: Cartas, Ministerios y Misiones Diplomaticas, AGN.

109 Admiral Rodgers to Sandford Fleming, 31 Oct. 1884, vol. 41, Fleming Papers. See also *Protocols of the Proceedings*, 207.

110 Bartky, *One Time Fits All*, 96. William Allen to Sandford Fleming, 18 Oct. 1887, vol. 1, Fleming Papers.

111 Bartky, *One Time Fits All*, 96.

112 Ibid., 95.

113 Admiral Rodgers to Sandford Fleming, 15 Dec. 1887, vol. 41, Fleming Papers. See also Message from the President to the House of Representatives Recommending taking action on the Prime Meridian Conference, 1888, vol. 122, Fleming Papers.

114 Sandford Fleming to Tondini de Quarenghi, Various Letters, vol. 13, Fleming Papers.

115 Bartky, *One Time Fits All*, 97.

116 *Daily News* (London), 31 Dec. 1884.

117 S.R. Franklin, General Order, 4 Dec. 1884, box 46, Newcomb Papers, LC.

118 Simon Newcomb to Nautical Almanac Office, 6 Dec. 1884, ibid.

119 S.R. Franklin to United States Naval Observatory, 11 Dec. 1884, RGOA, RGO7.146, CUL.

120 S.R. Franklin to William Christie, 10 and 15 Dec. 1884, Letters Received, PC42, entry 7, box 49A, Records of the United States Naval Observatory, RG 78, NA-USA.

121 S.R. Franklin to Simon Newcomb, 2 Jan. 1885, box 46, Newcomb Papers, LC.

122 S.R. Franklin, 31 Dec. 1884, Letters Received, PC42, entry 7, box 49A, Records of the United States Naval Observatory, RG 78, NA-USA.

123 Kikuchi Dairoku to J.C. Adams, 12 Dec. 1884, 24/16/2, Adams Papers.

124 William Christie to A.[dolphe] Hirsch, 5 Feb. 1885, RGOA, RGO7.148, CUL.

125 Third Report [1st Amendment] of the Joint Committee of the Canadian Institute and the Astronomical and Physical Society of Toronto on the Unification of Astronomical, Nautical, and Civil Time, Canadian Institute Papers, file 4-0-8, F1052, AO.

126 Ibid.

127 Struve, "The Resolutions of the Washington Meridian Conference," in Fleming, *Universal or Cosmic Time*, 1885, YA.2003.A.17994, BL.

CHAPTER FOUR

1 See Ogle, *The Global Transformation of Time*.

2 Darnton, *The Great Cat Massacre*, 5.

3 "A Mixed Timekeeper," *Aberdeen Weekly Journal*, 2 April 1887.

4 "Johnathan's Jokes," *Hampshire Telegraph and Sussex Chronicle*, 11 July 1885.

5 "What Time Is It?," *Manchester Courier and Lancashire General Advertiser*, 14 March 1895.

6 *Manchester Times*, 28 Sept. 1889.

7 *Observatory*, Nov. 1908.

8 Ibid.

9 *Punch* (London), 13 Dec. 1884.

10 Quoted in Galison, *Einstein's Clocks*, 122.

11 Ibid., 122.

12 *Observatory*, 1910, 188.

13 Duffy, "The Eight Hours Day Movement in Britain."

14 Barrows, *The Cosmic Time of Empire*, 87.

15 These are numerous accounts of the attack in RGOA, RGO7.58, CUL.

16 Astronomer Royal to the Secretary of the Admiralty, 24 Feb. 1913, ibid., RGO7.52, CUL.

17 The officers stayed long enough to form relationships with the observatory staff – one of them liked to tease the servants about keeping the noise down when the astronomer royal's infant children were crying. Wilson, *Ninth Astronomer Royal: The Life of Frank Watson Dyson*, 164; Frank Dyson to Scott, 10 Nov. 1915, RGOA, RGO7.52, CUL.

18 F.A.P. Barnard to Sandford Fleming, 2 Oct. 1884, vol. 3, Fleming Papers.

19 *Glasgow Herald*, 28 Jan. 1848.

20 Ibid.

21 *Hampshire Telegraph and Sussex Chronicle*, 25 Sept. 1852.

22 Howse, *Greenwich Time*, 109.

23 *Blackburn Standard*, 26 July 1848.

24 *Liverpool Mercury*, 17 Oct. 1848.

25 For more examples, see Howse, *Greenwich Time*, 105–13.

26 *North Wales Chronicle*, 18 Jan. 1848.

27 *Friday London Gazette*, 16 Aug. 1851.

28 Howse, *Greenwich Time*, 105.

29 Ibid., 105.

30 Greenwich's position of authority was a carefully constructed one. Scientific institutions rely on creating a perception of trust, credibility, and authority in order for their work to be transported and used. For more on how and why these qualities are precious in science, see Shapin, "Placing the View from Nowhere," 5–12; Latour and Woolgar, *Laboratory Life*, 187–208; and Johnston, "Managing the Observatory," 155–75.

31 *Birmingham Daily Post*, 27 Dec. 1884; *Daily News* (London), 18 Dec., 31 Dec. 1884.

32 *Daily News*, 31 Dec. 1884.

33 Ibid., 6 Feb. 1885; William Christie to Dr. Schram, 16 April 1891, RGOA, RGO7.146, CUL. W.H. Le Fevre, "A Standard of Time for the World: Address Delivered before the Balloon Society," 11 Sept. 1891, vol. 123, Fleming Papers.

34 "Juvenile Lectures," *Journal of the Society of Arts* 33, no. 1673 (Friday 12 Dec. 1884): 81.

35 "Juvenile Lectures," ibid., no. 1674 (Friday 19 Dec. 1884).

36 Norman Lockyer, "Universal Time: Our Future Clocks and Watches," ibid., no. 1677 (Friday 9 Jan. 1885): 172. Lockyer may have intended this lecture to occur on Christmas Eve. It is also possible that he meant New Year's Eve, as in the late nineteenth century traditions concerning Santa Claus involved variously the nights of 5, 24, and 31 December, hence Lockyer's gift-toting saint on New Year's Eve.

37 Ibid., 174.

38 Ibid., 186.

39 Ibid., 187.

40 Ibid., 187.

41 Ibid., 188.

42 Ibid., 186.

43 "Jottings," *Horological Journal* 27, no. 318 (Feb. 1885): 78; "Twenty-four O'Clock," ibid., no. 319 (March 1885): 90.

44 Sanders, "Beckett, Edmund."

45 "Twenty-four O'Clock," 90.

46 Ibid., 91.

47 "Jottings," *Horological Journal* 27, no. 318 (Feb. 1885): 78.

48 Ibid. .

49 Ibid.

50 Ibid., 79.

51 "New Standards of Time in the United States," *Horological Journal* 26, no. 304 (Dec. 1883): 53–4.

52 "Jottings," *Horological Journal* 27, no. 315 (Nov. 1884): 39.

53 Thomas Wright, "Bracebridge's Local and Mean Time Watch," ibid., no. 316 (Dec. 1884): 45.

54 E. Storer, "Twenty-four-Hour Dials," ibid., no. 317 (Jan. 1885): 69.

55 "Jottings," ibid., 66.

56 "Division of the Day in Southern Italy," *Horological Journal* 37, no. 318 (Feb. 1885): 75.

57 Ibid., .

58 "Jottings," *Horological Journal* 27, no. 319 (March 1885): 78–87.

59 J. Haswell, "Twenty-four-Hour Dials for Watches," ibid., no. 317 (Jan, 1885): 63.

60 "Jottings," ibid., 65.

61 "The New Time-o'-Day," ibid., 70.

62 "Jottings," ibid., no. 319 (March 1885): 88.

63 Ibid., 88.

64 Ibid.

65 Ibid.

66 "Combined Twelve and Twenty-four Hour Watch," *Horological Journal* 27, no. 318 (Feb. 1885): 75–6.

67 "Jottings," ibid., no. 319 (March 1885): 88.

68 Ibid..

69 Landes, *Revolutions in Time*, 287–90. See also *Horological Journal* 27, no. 318 (Feb. 1885): 77.

70 "Depression of the Watch Trade," *Horological Journal* 28, no. 331 (March 1886): 97.

71 Ibid.

72 *Horological Journal* 27, no. 318 (Feb. 1885): 77.

73 Ibid.

74 Ibid.

75 *Horological Journal* 28, no. 331 (March 1886): 103.

76 An 1895 editorial declared watchmaking "A Dead Industry." *Horological Journal* (Nov. 1895): 31.

77 "Automatic 24-Hour Dial," *Horological Journal* 27, no. 323 (July 1885): 149.
78 *Vo Key's Royal Pocket Index Key to Universal Time*, 8560.A.47, BL.
79 *Universal Lamp Time Chart*, 1898, 74.1865.C.18, BL.
80 *Martin's Tables*, 177–86, 223–33.
81 Ellis, "Description of the Greenwich Time Signal," 7.
82 Ibid., 10–11.
83 Advertisements of the rating competitions were published in the *Horological Journal*.
84 Engineer in Chief R.S. Culley to Frank Scudamore, 10 Nov. 1870, Time Signals, Post Office Papers (hereinafter POP), Post 30.2536, BT.
85 Ibid.
86 Ibid.
87 Ibid.
88 Frank Scudamore, Response to Engineer in Chief, 10 Nov. 1870, Time Signals, POP, Post 30.2536, BT.
89 J.B. Pearson to Post Office Secretary, 25 April 1881, Time Signals, ibid.
90 Ibid.
91 Postmaster General to J.H. T[*illegible*] Postmaster Cambridge, 29 April 1881, Time Signals, POP, Post 30.2536, BT.
92 Postmaster General Instructions on J.B. Pearson Request, 7 May 1881, ibid.; Post Office to J.B. Pearson, 12 May 1881, ibid.
93 Post Office to J.B. Pearson, 12 May 1881, ibid.
94 J.B. Pearson to H. Fawcett, Postmaster General, 28 July 1881, ibid.
95 J.B. Pearson to H. Fawcett, Postmaster General, 28 July 1881, ibid.
96 J.B. Pearson to H. Fawcett, Postmaster General, 28 July 1881, ibid.
97 H. Fawcett to J.B. Pearson, 2 Aug. 1881, ibid.
98 Ibid.
99 J.B. Pearson to H. Fawcett, 11 Aug. 1881, Time Signals, POP, Post 30.2535, BT.
100 H. Darwin to H. Fawcett, 31 Dec. 1881, Time Signals, POP, Post 30.2536, BT.
101 Ibid.
102 H. Fawcett, Note Concerning Reply to H. Darwin, 12 Jan. 1882, ibid.
103 Edward Graves, Engineer in Chief, to H. Fawcett, 10 Feb. 1882, ibid.
104 Submits Report of Engineer in Chief on cost of time signals to Postmaster General, 21 Feb. 1882, ibid.
105 See William Christie, *Report of the Astronomer Royal to the Board of Visitors of the Royal Observatory*, 2 June 1888, 20, RGOA, RGO17.1.4, CUL.
106 William Christie to William Preece, 9 July 1888, ibid., RGO7.254, CUL.
107 Christie, *Report*, 2 June 1888, 20.
108 Ibid.

109 William Christie to the Secretary of the Admiralty, 27 June 1887, Greenwich Observatory, POP, Post 30.523C, BT.

110 William Preece to William Christie, 13 June 1888, RGOA, RGO7.254, CUL.

111 William Christie to William Preece, 20 June 1888, Greenwich Observatory, POP, Post 30.523C, BT.

112 William Christie to William Preece, 9 July 1888, RGOA, RGO7.254, CUL.

113 William Preece to William Christie, 10 July 1888, ibid.

114 William Christie to Henry Cecil Raikes, 12 July 1888, ibid. A copy can also be found in Greenwich Observatory, POP, Post 30.523C, BT.

115 Henry Cecil Raikes to the Secretary of the Admiralty, 10 July 1888, Greenwich Observatory, POP, Post 30.523C, BT.

116 Ibid.

117 Henry Cecil Raikes to the Secretary of the Admiralty, 23 July 1888 [see also Post Office Memo, CABP to Raikes, 19 July 1888], Greenwich Observatory, POP, Post 30.523C, BT.

118 Post Office Memo, CABP to Henry Cecil Raikes, 19 July 1888, ibid.

119 Lords Commissioners of the Admiralty to William Christie, 7 Aug. 1888, RGOA, RGO7.254, CUL.

120 William Christie to the Secretary of the Admiralty, 11 Aug. 1888, ibid.

121 Henry Cecil Raikes to the Secretary of the Admiralty, 11 Sept. 1888, Greenwich Observatory, POP, Post 30.523C, BT.

122 Evan Macgregor to Henry Cecil Raikes, 1 Oct. 1888, ibid.

123 Lords Commissioners of the Admiralty to William Christie, 1 Oct. 1888, RGOA, RGO7.254, CUL.

124 Ibid.

125 William Christie to the Secretary of the Admiralty, 12 Oct. 1888, ibid.

126 Ibid.

127 Ibid.

128 Ibid.

129 Ibid.

130 Lords Commissioners of the Admiralty to William Christie, 1 Dec. 1888, RGOA, RGO7.254, CUL.

131 Ibid.

132 Mary Brück, "Lady Computers," 85.

133 Higgitt, "A British National Observatory," 621, note 51. As Higgitt comments, this solution was a mere "stopgap."

134 William Christie, *Report of the Astronomer Royal to the Board of Visitors of the Royal Observatory,* 1 June 1889, 21–22, RGOA, RGO17.1.4, CUL.

135 William Christie, *Report of the Astronomer Royal to the Board of Visitors of the Royal Observatory,* 6 June 1891, 19, ibid.

136 William Christie, *Report of the Astronomer Royal to the Board of Visitors of the Royal Observatory,* 4 June 1892, 23, ibid.

137 Post Office Telegraphs Articles of Agreement, 31 March 1884, Synchronisation Etc. of Clocks by Electric Current, POP, Post 30.531, BT.

138 Ibid.

139 Rooney, *Ruth Belville,* 62.

140 See Synchronisation Etc.

141 Question of Infringement of Postmaster General's Rights, Counsel's Opinion and Opinion of the Law Officers of the Crown, 1888, Synchronisation Etc.

142 Ibid.

143 Greenwich Time Company Advertisement, 1911, Synchronisation Etc.

144 See Silvanus Philips Thompson Collection, SPT/65, IET.

145 Magneta Company Advertisement, 1906, ibid.

146 Synchronome Company Advertisement, undated, ibid.

147 Ibid.

148 Greenwich Time Company Advertisement, 1911, Synchronisation Etc.

149 Ibid.

150 *Advertising News,* 2 June 1905, Silvanus Philips Thompson Collection, SPT/65, IET.

151 Magneta Company Advertisement, 1906, ibid.; Some Recent Installations, ibid.

152 Electric Time Service, Synchronome Company Advertisement, undated, ibid.

153 Ibid.; Some Extensive Premises in Which Synchronome Time-Services are Established, undated, Silvanus Philips Thompson Collection, SPT/65, IET.

154 "The House that Jack Built," Synchronome Company Advertisement, undated, ibid.

155 Ibid.

156 Ibid.

157 Ibid.

158 Standard Time Company to the General Post Office, 2 May 1884, Synchronisation Etc.

159 District Manager of Telephones, S.E. Lanc, to Post Office Secretary, 15 July 1914, ibid.

160 Secretary of the Oldham Master Cotton Spinners' Association to the District Manager of Telephones, S.E. Lanc, 4 Dec. 1913, ibid.

161 Post Office Secretary to District Manager of Telephones, S.E. Lanc, 25 July 1914, ibid.

162 Post Office Memo for the Postmaster General, 4 Jan. 1913, Synchronisation of Clocks, POP, Post 30.2042B, BT.

163 *Daily Express* (London), 26 April 1913.

164 British Science Guild, Report of Committee on the Subject of Synchronisation of Clocks in London, and in other parts of Great Britain, 1 Aug. 1908, Synchronisation of Clocks.

165 Viator, "Public Clocks," *Horological Journal* 27, no. 324 (Aug. 1885): 171.

166 "Lying Clocks," *The Times*, Jan. 1908, Time Signals, POP, Post 30.2536, BT.

167 See Minutes of the United Wards Club, 4 March 1908, United Wards Club Papers, vol. 1, Ms. 11723, LMA; Minutes of the Committee of the United Wards Club, 19 Feb. 1908, ibid., vol. 2, Ms. 11724, LMA; "The Time of a Great City," *Transactions of the United Wards Club* no. 81 (13 March 1908): 1–4, ibid., vol. 1, Ms., 21483, LMA; St John Winne, "The Time of a Great City: A Plea for Uniformity" (4 March 1908), Silvanus Philips Thompson Collection, SPT/65, IET; Rooney, *Ruth Belville*.

168 Winne, "The Time of a Great City," 9.

169 Ibid., 23.

170 Ibid 23–4. By this he meant that Dent and Co. had built the main clock in Greenwich Observatory, which the Post Office used to distribute the time to the STC.

171 "Lady Who Conveys the Time," untitled paper clipping, 1908, RGOA, RGO7.96, CUL.

172 For ease of understanding, I sometimes use their first names to distinguish mother from daughter.

173 Mrs E. Henry Belville (Maria) to George Airy, 31 Aug. 1856, RGOA, RGO6.4, CUL.

174 Ruth Belville, "Some Account of John Henry Belville and the Distribution of G.M.T. to Chronometer Makers in London," 1938, RGOA, RGO74.6.2, CUL.

175 Ruth Belville to Mr Lewis, 22 Jan. 1910, RGOA, RGO7.96, CUL; "Selling the Time to London," *Evening News* (London), 3 April 1929.

176 "Selling the Time to London."

177 Ruth Belville, "History of the Belville Time Service," untitled and undated (possibly 1916), RGOA, RGO7.96, CUL.

178 Rooney, *Ruth Belville*, 62.

179 "Telling the Time as an Occupation," *Daily News and Leader*, 29 April 1913. Rooney, *Ruth Belville*.

180 Entries for Maria Belville in *Census Returns of England and Wales, 1841, 1861, 1871, 1891*, NA-UK; entries for Ruth Belville in *Census Returns of England and Wales, 1861, 1871, 1881, 1891, 1901, 1911*, NA-UK.

181 Entry for Ruth Belville in *Census Returns of England and Wales, 1901*, NA-UK.

182 Mrs E. Henry Belville (Maria) to George Airy, 6 Aug. 1856, RGOA, RGO6.4, CUL.

183 George Airy to Maria Belville, 11 Aug. 1856, ibid.

184 Mrs E. Henry Belville (Maria) to George Airy, 6 Aug. and 21 Aug. 1856, ibid.

185 The Admiralty to George Airy, 3 Sept. 1856, ibid.; George Airy to Maria Belville, 4 Sept. 1856, ibid.

186 Mrs E. Henry Belville (Maria) to George Airy, 31 Aug. 1856, ibid. The Belvilles' "regulator" was a chronometer made by John Arnold in 1794 (no. 485). It currently resides in the Clockmaker's Museum at the Science Museum, South Kensington, London.

187 George Airy to Maria Belville, 3 Nov. 1856, RGOA, RGO6.43, CUL.

188 Mrs E. Henry Belville (Maria) to George Airy, 3 Nov. 1856, RGOA, RGO6.4, CUL.

189 George Airy to Maria Belville, 3 Nov. 1856, RGOA, RGO6.43, CUL.

190 Ruth Belville to William Christie, 10 June 1892, RGOA, RGO7.254, CUL.

191 H. Turner to Ruth Belville, 12 June 1892, ibid.

192 "Greenwich Mean Time," *Daily Graphic* (London), 31 Oct. 1892.

193 H.H. Turner, "Greenwich Mean Time," ibid., 1 Nov. 1892.

194 "Woman Who Sold the Time," *Daily Express* (London), 7 March 1908.

195 "Woman Who Sells the Time: Strange Profession of the Belville Family," ibid., 9 March 1908.

196 "Greenwich Clock Lady: Romance of a Regular Visitor to the Observatory," *Kentish Mercury*, 13 March 1908.

197 "Lady Who Has Inherited a Strange Trade," *Daily News and Leader* (London), 29 April 1913.

198 "The Belville Tradition," *Observer* (London), 24 Aug. 1913.

199 "Maidenhead Lady Who Distributes the Time: A Unique Position," *Maidenhead Advertiser*, 11 March 1908.

200 Ruth Belville, "The Belville Tradition," *Observer*, 31 Aug. 1913.

201 Belville, "History of the Belville Time Service."

202 Rooney, *Ruth Belville*, 61, 146, 173.

203 Rooney and Nye, "'Greenwich Observatory Time,'" 29; Rooney, *Ruth Belville*.

204 Rooney, *Ruth Belville*, 146, 173.

205 William Christie, *Report of the Astronomer Royal to the Board of Visitors of the Royal Observatory*, 5 June 1880, 17–18, RGOA, RGO17.1.4, BL.

206 See RGOA, RGO7.253, CUL.

207 See A12/54 and A14/65, Smyth Papers.

208 See RGOA, RGO15.107 and RGO15.108, CUL.

209 J.B. Chapman to Mr Graves, 5 Aug. 1887, Greenwich Observatory, POP, Post 30.523C, BT.

210 Post Office Memorandum, 28 April 1915, Synchronisation Etc.

CHAPTER FIVE

1 Ryerson has a mixed legacy. He advocated for progressive policies regarding free, compulsory education and the separation of church and state in schools, but he was also one of the architects of the devastatingly harmful residential school system for Indigenous children.

2 Oreopoulos, "Canadian Compulsory School Laws," 8–9. And for Quebec, see Fateux, "Quebec," at http://faculty.marianopolis.edu/c.belanger/quebechistory/encyclopedia/QuebecProvinceof.htm (accessed 24 June 2019).

3 Baldus and Kassam, "'Make Me Truthful, Good, and Mild.'" See also Prentice and Houston, eds., Family, School, and Society, 178–82, 281.

4 Willinski, Learning to Divide the World, 2–3.

5 Rural areas were also the most hesitant to adopt daylight saving time when that innovation gained popularity a few decades later. See Rudy, "Do You Have the Time?"

6 Anna Molander to Greenwich Observatory, 23 June 1909, RGOA, RGO7.140, CUL.

7 William Cranch Bond at Harvard University established the first direct connection to Greenwich time in North America. Brooks, "Time," 184–85.

8 U.S. Naval Observatory to John White, 4 Feb. 1882, Superintendent's Office Outgoing Correspondence, box 1, Records of the U.S. Naval Observatory, LC.

9 Stephens, "Before Standard Time," 116. The Dudley Observatory, not related to a university at the time, also had a time service.

10 Stephens, "Before Standard Time," 117.

11 William Allen to Superintendent of the Naval Observatory, 6 Oct. 1883, Letters Received, PC42, entry 7, box 49A, Records of the United States Naval Observatory, RG 78, NA-USA.

12 Robert Wilson Shufeldt to Edwin Leigh, 15 Oct. 1883, Miscellaneous Letters Sent, PC42, entry 4, vol. 5, ibid.

13 Ibid.

14. Ibid.

15 Ibid.

16 Ibid.

17 Henry Pritchitt to Robert Shufeldt, 16 Oct. 1883, Letters Received, PC42, entry 7, box 49A, Records of the United States Naval Observatory, RG 78, NA-USA..

18 Ibid.

19 Ibid.

20 Ibid.

21 Robert Shufeldt to Henry Pritchitt, 18 Oct. 1883, Miscellaneous Letters Sent, PC42, entry 4, vol. 5, Records of the United States Naval Observatory, RG 78, NA-USA.

22 Henry Pritchitt to Robert Shufeldt, 26 Oct. 1883, Letters Received, PC42, entry 7, box 49A, ibid.

23 Ibid.

24 Robert Shufeldt to Henry Pritchitt, 26 Oct. 1883, Miscellaneous Letters Sent, PC42, entry 4, vol. 5, Records of the United States Naval Observatory, RG 78, NA-USA.

25 Robert Shufeldt to Henry Pritchitt, 29 Oct. 1883, ibid.

26 Bartky, *Selling the True Time*, 181.

27 Ibid., 187–8, 199–200.

28 Ibid., 211–12.

29 Thomson, *The Beginning of the Long Dash*, xii. For more on the Toronto Meteorological Observatory's time responsibilities, see Thomas, *The Beginnings of Canadian Meteorology*.

30 Brooks, "Time,," 181.

31 Annual Report of Charles Smallwood, Director of the Montreal Observatory, 1870, RG 6, A1, vol. 10, LAC.

32 Memorandum on the transfer of the Toronto Observatory from the Marine to Interior Department, 1892, RG 15, vol. 669, LAC.

33 Stewart, "The Time Service," 97, 99, 102.

34 B.C. Webber to Astronomer Royal, Greenwich, 7 Sept. 1909, RGOA, RGO7.252, CUL.

35 The Astronomer Royal, Cambridge, to the Director of the Meteorological Office, Toronto, 26 Nov. 1909, ibid.

36 Edward David Ashe to the Governor General, 19 Nov. 1856, RG 93, vol. 82, Quebec Observatory, LAC.

37 Petition of the Canadian Institute to Erect and Endow a 'National Canadian Astronomical Observatory,' 18 March 1857, ibid.

38 William Ellis to William Ashe, 16 April 1869, ibid.; R. Stupart to Arthur Smith, 9 Jan. 1895, ibid.

39 Andrew Gordon to William Ashe, 16 Feb. 1888, ibid.

40 See Edward Ashe to Mr Smith, 20 Oct. 1879, ibid. William Ashe to P. Garneau, 16 May 1888, ibid.

41 Deputy Minister of Marine, Memo, 31 Dec, 1889, RG 12, vol. 1231, Meteorological Service, LAC.

42 Chief Signal Officer to William Smith, 17 Jan. 1890, ibid.

43 Evan MacGregor to the Board of Trade, 25 Jan. 1890, ibid.

44 Memo to Minister, 17 Feb. 1890, ibid.

45 Charles Carpmael to William Smith, 12 March 1890, ibid.

46 Charles Hosmer to Charles Carpmael, 27 March 1890, ibid.

47 H.P. Dwight to Charles Carpmael, 28 March 1890, ibid.

48 William Ashe to Charles Carpmael, 18 July 1891, ibid.

49 Ibid.

50 Memorandum on the Time Ball at the Citadel, Quebec, 29 July 1891, ibid.

51 Charles Carpmael to unknown, 17 Aug, 1891, ibid.

52 R. Stupart to Arthur Smith, 7 July 1894, RG 93, vol, 82, Quebec Observatory, LAC.

53 *Massey's Illustrated,* Jan. 1884.

54 *Globe* (Toronto), 22 Oct. 1883.

55 Ibid.

56 Ibid.

57 Ibid.

58 Ibid.

59 Ibid.

60 Ibid.

61 Ibid.

62 *Globe,* 13 Nov. 1883.

63 Ibid., 26 Oct. 1883. According to Gordon, the edge of the time zone was marked by 82.30 degrees west. Sarnia's longitude lay just over the line at 82.40 degrees west.

64 Ibid., 21 Nov. 1883.

65. Ibid., 22 Nov. 1883.

66 *Truth,* 24 Nov. 1883.

67 *Varsity,* 17 Nov. 1883.

68 Bloomfield Douglas to the Director of the Quebec Observatory, 6 May 1895, RG 93, vol. 82, Quebec Observatory, LAC.

69 G.W. Wicksteed to Sandford Fleming, 7 Dec. 1883, vol. 53, Fleming Papers.

70 G.W. Wicksteed to Sandford Fleming, 13 Dec. 1883, ibid.

71 G.W. Wicksteed to Sandford Fleming, 22 Dec. 1883, ibid.

72 Ironically in a letter about keeping time correctly, the letter is dated 31 February (!) 1885. G.W. Wicksteed to Sandford Fleming, 31 Feb. 1885, ibid.

73 Ibid.

74 Ibid.

75 Thomson, *The Beginning of the Long Dash,* 34.

76 *Ottawa Free Press,* 31 May 1893, RG 12, vol. 1220, Meteorological Service, LAC. Fleming used this case to push for a legal time, to avoid more confusion.

77 Ibid.

78 *Ottawa Citizen,* undated, vol. 53, 20651, Fleming Papers.

79 Ibid.

80 *Legal News* 8, no. 15, 11 April 1885.

81 *Quebec Daily Mercury*, 20 Nov. 1883.
82 Thomson, *The Beginning of the Long Dash,* 34. See also RG 12, vol. 1220, Meteorological Service, LAC; vol. 47, 50, Fleming Papers.
83 D.R. Cameron to W. Smith, 30 Nov. 1891, RG 12, vol. 1220, Meteorological Service, LAC.
84 Sandford Fleming to William Smith, 14 May 1892, ibid.
85 Sandford Fleming to William Smith, 21 March 1892, ibid.
86 D.R. Cameron to William Smith, 7 March 1892, ibid.
87 Thomson, *The Beginning of the Long Dash,* 27.
88 Charles Carpmael to William Smith, 20 Feb., 2 March 1892, RG 12, vol. 1220, Meteorological Service, LAC.
89 Canadian Institute Memorial, The Uniform Notation of Time by all Nations, 1888, Canadian Institute Papers, file 4-0-3, F1052, AO.
90 F.A.P. Barnard to J.K. Rees, 4 Feb. 1888, vol. 40, Fleming Papers.
91 William Allen to J.K. Rees, 9 Jan, 1888, ibid.
92 Canadian Institute Memorial, The Uniform Notation.
93 Governor General to Charles Carpmael, undated, Canadian Institute Papers, file 4-0-3, F1052, AO.
94 Lord Knutsford to the Governor General, 23 July 1888, ibid. For the Netherlands, see British Legation, The Hague, to the Canadian Institute, 3 July 1888, ibid.
95 Acting President, Orange Free State, to Sir Hercules Robinson, 17 Sept. 1888, ibid.
96 Mr Kennedy to the Marquis of Salisbury, 8 July 1888, ibid.; India Office to Colonial Office, 4 July 1888, ibid.; Colonial Secretary to the Canadian Institute, 2 Nov. 1888, ibid.
97 Education Department of Ontario to Charles Carpmael, 27 March 1888, ibid; Canadian Institute to the Chief Superintendent of Education, Ontario, 15 May 1888, ibid.
98 Chief Superintendent of Education, New Brunswick, to Charles Carpmael, 4 June 1888, ibid.; North West Territories Board of Education to the Canadian Institute, 9 Aug. 1888, ibid.
99 Département de l'Instruction Publique, Section Catholique, to Canadian Institute, 2 June 1888, ibid.
100 Superintendent of Schools, Manitoba, to the Canadian Institute, 4 June 1888, ibid.
101 David Allison to Charles Carpmael, 6 June 1888, ibid.; D. Montgomery to Charles Carpmael, 9 June 1888, ibid.; S.D. Pope to R.W. Young, 13 June 1888, ibid.; Quebec Department of Public Instruction to Charles Carpmael, 26 June 1888, ibid.
102 L. McCaskey to Sandford Fleming, undated, ibid.

103 L. McCaskey, Longitude and Geographical Time Chart, ibid.

104 Ibid.

105 Ibid.

106 *Education Weekly*, 1 Jan. 1885.

107 *Home and School Supplement*, Sept. 1885; see also March 1887 and December 1886.

108 Ibid., March 1886.

109 E. Peel to W. Whyte, 11 Dec. 1886, vol. 51, Fleming Papers; [*illegible*] to W. Whyte, 11 De. 1886, ibid.

110 Mr McAdam to W. Whyte, 11 Dec. 1886, ibid.

111 July Proceedings, Report of the Special Committee on Standard Time, vol. 2, Fleming Papers.

112 Fred Brooks to the American Society of Civil Engineers, 3 Nov. 1888, ibid.

113 Fred Brooks to the American Society of Civil Engineers, 3 Nov. 1888, ibid.

114 Archibald, "Resistance to an Unremitting Process," 107.

115 Miller, *Shingwauk's Vision*, 10.

116 Ibid., 11.

117 Ibid., 193.

118 Similar practices occurred elsewhere in the empire too. Giordanno Nanni, for example, has shown how Greenwich's constructed authority became a tool of colonization through bells and schedules in South African schools, enforcing structure and undermining local timekeeping practices. Nanni, *The Colonisation of Time*, 191–212.

119 Smith, *Sacred Feathers*, 160, 226.

120 "Rules and Regulations of the Mississaugas of the Credit," 16, 19.

121 Ibid., 16.

122 Ibid., 16.

123 Inspector's Report on the Moravian 'Reserve' Indian School, June 1885, RG 10, vol. 5991, LAC.

124 Inspector's Report on the Moravian 'Reserve' Indian School, Feb. 1895, ibid.

125 Inspector's Report on the Moravian 'Reserve' Indian School, Jan. 1893, ibid.

126 Inspector's Report on the Moravian 'Reserve' Indian School, 1891, ibid.

127 Inspector's Report on the Moravian 'Reserve' Indian School, March 1886, ibid.

128 Smith, *Sacred Feathers*, 239, 244.

129 *Indian*, every issue.

130 Ibid., 12 May 1886.

131 Ibid.

132 Ibid., 9 June 1886.

133 Ibid.

134 Johannah Hill in Ontario, Canada, Marriages, 1801–1928, 1933–1934, Registrations of Marriages, 1869–1928, Series Ms. 932, reel 46, AO.

135 *Indian,* 7 July 1886.
136 Inspector's Report on the Moravian 'Reserve' Indian School, 1888, RG 10, vol, 5991, LAC.

CONCLUSION

1 Knorr-Cetina, *The Manufacture of Knowledge,* 5.
2 Ogle, *The Global Transformation of Time.*
3 E.P. Thompson, "Time."
4 Ibid., 59–60.
5 Ibid. 60.
6 Ibid.
7 Ibid., 61.
8 Ibid., 84.
9 Ibid., 80.
10 Glennie and Thrift, *Shaping the Day,* 132.
11 Landes, *Revolutions in Time,* 58; Glennie and Thrift, *Shaping the Day,* 40.
12 Ogle, *The Global Transformation of Time,* 47–9, 55–6, 62–4, 71–2.
13 Stephen Kern makes a similar case for a backlash to modernizing timekeeping. Kern, *The Culture of Time and Space.*

Bibliography

ARCHIVAL SOURCES

Archives of Ontario (AO), Toronto
  Canadian Institute Papers, F1052.
  Registrations of Marriages, 1869–1928, Series Ms. 932, Reel 46.
Archives of the Royal Society (ARS), London
  Royal Society Council Minutes, CM O17.
Archivo General de la Nacion (AGN), Santo Domingo, Dominican Republic
  Manuel de Jesús Galvan, vol. LIX, Textos Reunidos 4: Cartas, Ministerios y
    Misiones Diplomaticas.
Bibliothèque de l'Institut de France (BIF), Paris
  Correspondance de Jules Janssen, Ms. 4133.
British Library (BL), London
  Fleming, Sandford. *Universal or Cosmic Time*, 1885, YA.2003.A.17994.
  Papers of Lt-Gen Sir Richard Strachey (Strachey Papers). Mss. Eur F127,
    vols. 114, 122, 132, 151, 187, 188.
  Snow, William Parker. "An International Prime Meridian: Greater Safety at
    Sea by Uniform Routes," Circular Letter, Mic.A.19863.
  Universal Lamp Time Chart, 1898, 74.1865.C.18.
  *Vo Key's Royal Pocket Index Key to Universal Time upon the Face of Every
    Watch and Clock* (London: Express Printing Co., 1901?), 8560.A.47.
BT Archives (BT = British Telecom), London
  Post Office Papers (POP), Post 30.523C, Post 30.531, Post 30.2042B, Post
    30.2536.
Cambridge Astronomy Library (CAL)
  Fleming, Sandford. *Time Reckoning*, 1879, Canadian and U.S. Papers on
    Time Reckoning, E.13.1.

Cambridge University Library (CUL)
   Royal Greenwich Observatory Archives (RGOA): RGO6.4, RGO6.13,
      RGO6.43, RGO6.365, RGO7.52, RGO7.58, RGO7.96, RGO7.138,
      RGO7.140, RGO7.142, RGO7.146, RGO7.147, RGO7.148, RGO7.252,
      RGO7.253, RGO7.254, RGO8.150, RGO9.625, RGO15.107, RGO15.108,
      RGO17.1.4, RGO74.6.2.
Institute of Engineering and Technology Archives (IET), London
   Silvanus Philips Thompson Collection, SPT/65.
Library and Archives Canada (LAC), Ottawa
   Fleming Papers, MG 29 B1, vols. 1, 2, 3, 7, 11, 13, 14, 27, 30, 33, 39, 40, 41,
      47, 50, 51, 53, 54, 63, 65, 81, 105, 122, 123.
   RG 6, RG 10, RG 12, RG 15, RG 30, RG 93.
Library of Congress (LC), Washington, DC
   Newcomb Papers, box 6, box 38, box 46.
   Superintendent's Office Outgoing Correspondence, box 1, The Records of
      the U.S. Naval Observatory.
London Metropolitan Archives (LMA), London, England
   United Wards Club Papers, Ms. 11723 vol. 1, Ms. 11724 vol. 2, Ms. 21483
      vol. 1.
National Archives of the United Kingdom (NA-UK), London
   Census Returns of England and Wales, 1841, 1861, 1871, 1881, 1891,
      1901, 1911.
   Colonial Office Papers, CO201.601, CO309.127, CO340.2, CO42.775,
      CO42.776, CO42.779.
   Foreign Office Papers, FO566.19, FO5.1872, FO5.1884, FO5.1886,
      FO5.1887.
National Archives of the United States of America (NA-USA), Washington, DC
   Despatches from U.S. Ministers to Great Britain 1791–1906, Microfilm
      M30 146.
   Domestic Letters of the Department of State 1784–1906, Microfilm M40 101.
   Letters Received, PC42, entry 7, box 49A, Records of the United States
      Naval Observatory, RG 78.
   Letters to the International Meridian Conference 1884, box 1, Records of
      International Conferences, Commissions, and Expositions, RG 43.
   Miscellaneous Letters Sent, PC42, entry 4, vol. 5, Records of the United
      States Naval Observatory, RG 78.
Royal Astronomical Society Archives (RAS), London
   Royal Astronomical Society Papers, Part 1.
Royal Geographical Society Archives (RGS), London
   RGS/CB6/1377.

ography234

Royal Observatory of Edinburgh (ROE)
Charles Piazzi Smyth Archives [on loan from the Royal Society of Edinburgh].
Charles Piazzi Smyth Papers, A12/52, A12/54, A12/55, A13/58, A13/59, A14/65, A14/66, A17/95.
St John's College Library, Cambridge (SJCL)
John Couch Adams Papers.

NEWSPAPERS

Aberdeen Weekly Journal, 1887.
Birmingham Daily Post, 1884.
Blackburn Standard, 1848.
Daily Express (London), 1908–13.
Daily Graphic (London), 1892–94.
Daily News (London), 1884–85.
Daily News and Leader (London), 1913.
Education Weekly, 1885.
Evening News (London), 1929.
Friday London Gazette, 1851.
Glasgow Herald, 1848.
Globe (Toronto), 1883.
Hampshire Telegraph and Sussex Chronicle, 1852–85.
Home and School Supplement, 1885–87.
Horological Journal, 1883–95.
Indian (Hagersville), 1886.
Indianapolis Sentinel, 1883.
Journal of the Society of Arts, 1884–85.
Kentish Mercury, 1908.
Legal News, 1885.
Liverpool Mercury, 1848.
Maidenhead Advertiser, 1908.
Manchester Courier and Lancashire General Advertiser, 1895.
Manchester Times, 1889.
Massey's Illustrated (Toronto), 1884.
New York Times, 1884.
North Wales Chronicle, 1848.
Observatory (London), 1908–10.
Observer (London), 1913.
Ottawa Citizen, c. 1885

Ottawa Free Press, 1893.
Punch (London), 1884.
Quebec Daily Mercury, 1883.
The Times (London), 1908.
Truth, 1883.
Varsity (University of Toronto), 1883.

SECONDARY SOURCES

Adas, Michael. *Machines as the Measure of Men: Science, Technology, and Ideologies of Western Dominance*. Ithaca, NY: Cornell University Press, 2015.
Allen, William. *Short History of Standard Time*. Philadelphia: Stephen Greene Printing Company, 1904.
Anderson, Martin. "The Development of British Tourism in Egypt, 1815 to 1850." *Journal of Tourism History* 4, no. 3 (2012): 259–79.
Archibald, Jo-Ann. "Resistance to an Unremitting Process: Racism, Curriculum, and Education in Western Canada." In *The Imperial Curriculum: Racial Images and Education in the British Colonial Experience*, ed. J.A. Mangan. London: Routledge, 1993.
Aubin, David. "On the Epistemic and Social Foundations of Mathematics as Tool and Instrument in Observatories, 1793–1846." In *Mathematics as a Tool: Tracing New Roles of Mathematics in the Sciences*, ed. Johannes Lenhard and Martin Carrier, 177–96. Cham, Switzerland: Springer International, 2017.
Baldus, Bernd, and Meenaz Kassam. "'Make Me Truthful, Good, and Mild': Values in Nineteenth-Century Ontario Schoolbooks." *Canadian Journal of Sociology* 21, no. 3 (1996): 327–58.
Barrows, Adam. *The Cosmic Time of Empire: Modern Britain and World Literature*. Berkeley: University of California Press, 2011.
Bartky, Ian. "The Adoption of Standard Time." *Technology and Culture* 30 (1989): 25–56.
– "Inventing, Introducing, and Objecting to Standard Time." *Vistas in Astronomy* 28 (1985): 105–11.
– *One Time Fits All: The Campaign for Global Uniformity*. Stanford, CA: Stanford University Press, 2007.
– "Sandford Fleming's First Essays on Time." NAWCC *Bulletin* 50, no. 1 (2008): 3–11.
– *Selling the True Time: Nineteenth Century Timekeeping in America*. Stanford, CA: Stanford University Press, 2000.

Blaise, Clark. *Time Lord: Sir Sandford Fleming and the Creation of Standard Time.* New York: Pantheon Books, 2000.

Brooks, Randall. "Time, Longitude Determination and Public Reliance upon Early Observatories." In *Profiles of Science and Society in the Maritimes prior to 1914,* ed. Paul Bogaard. Sackville, NS: Acadiensis Press, 1990.

Brück, Hermann, and Mary Brück. *The Peripatetic Astronomer: The Life of Charles Piazzi Smyth.* Bristol: A. Hilger, 1988.

Brück, Mary. "Lady Computers at Greenwich in the Early 1890s." *Quarterly Journal of the Royal Astronomical Society* 36 (1995): 83–95.

Burpee, Lawrence. *Sandford Fleming: Empire Builder.* London: Oxford University Press, 1915.

Burton, Antoinette. *At the Heart of the Empire: Indians and the Colonial Encounter in Late-Victorian Britain.* Berkeley: University of California Press, 1998.

Butler, Judith. *Gender Trouble: Feminism and the Subversion of Identity.* New York: Routledge, 1990.

Chapman, Allan. "Sir George Airy (1801–1892) and the Concept of International Standards in Science, Timekeeping, and Navigation." *Vistas in Astronomy* 28 (1985): 321–8.

Clark, Anna. *The Struggle for the Breeches: Gender and the Making of the British Working Class.* Berkeley: University of California Press, 1997.

Clark, Josiah Latimer. *Transit Tables for 1884: Giving the Greenwich Mean Time of the Transit of the Sun, and of Certain Stars, for Every Day in the Year, etc.* London: E. & F.N. Spon, 1884.

Cooper, Frederick. *Colonialism in Question: Theory, Knowledge, History.* Berkeley: University of California Press, 2005.

Creet, Mario. "Sandford Fleming and Universal Time." *Scientia Canadensis* 14, nos. 1–2 (1990): 66–89.

Darnton, Robert. *The Great Cat Massacre and Other Episodes in French Cultural History.* New York: Basic Books, 1984.

Davidoff, Leonore, and Catherine Hall. *Family Fortunes: Men and Women of the English Middle Class 1780–1850.* Chicago: University of Chicago Press, 1991.

Deringil, Selim. *Conversion and Apostasy in the Late Ottoman Empire.* Cambridge: Cambridge University Press, 2012.

Dolan, Graham. "Christie's Lady Computers – The Astrographic Pioneers of Greenwich." In *The Royal Observatory Greenwich,* at http://www.royal observatorygreenwich.org/articles.php?article=1280 (accessed 1 July 2020).

Draper, John. *History of the Conflict between Religion and Science.* New York: D. Appleton, 1874.

Drayton, Richard. *Nature's Government: Science, Imperial Britain, and the 'Improvement' of the World*. New Haven, CT: Yale University Press, 2000.

Duffy, A.E.P. "The Eight Hours Day Movement in Britain, 1886–1893." *Manchester School* 36, no. 3 (2008): 203–22.

Edney, Matthew. *Mapping an Empire: The Geographical Construction of British India, 1765–1843*. Chicago: University of Chicago Press, 1997.

Ellis, William. "Description of the Greenwich Time Signal." *Greenwich Astronomical Observations* (1879): J1–J13.

Endersby, Jim. *Imperial Nature: Joseph Hooker and the Practices of Victorian Science*. Chicago: University of Chicago Press, 2008.

Fateux, Aegidius. "Quebec." *The Quebec History Encyclopedia*, ed. Claude Belanger. http://faculty.marianopolis.edu/c.belanger/quebechistory/encyclopedia/QuebecProvinceof.htm (accessed 24 June 2019).

Fisher, Michael. *Counterflows to Colonialism: Indian Travellers and Settlers in Britain, 1600–1857*. Delhi: Permanent Black, 2004.

Fleming, Sandford. *The Adoption of a Prime Meridian to Be Common to All Nations*. 1881. https://archive.org/details/cihm_03125/page/n3/mode/2up (accessed 5 March 2020).

– *The International Prime Meridian Conference: Recommendations Suggested*. 1884, at https://archive.org/details/cihm_03131/page/n4/mode/2up (accessed 5 May 2018).

– *Uniform Non-Local Time (Terrestrial Time)*. Self-published, 1878, at https://archive.org/stream/cihm_03138#page/n11/mode/2up (accessed 12 September 2020).

Forbes, George. *The Transit of Venus*. London and New York: MacMillan: 1874.

France, Peter. *The Rape of Egypt: How the Europeans Stripped Egypt of Its Heritage*. London: Barrie & Jenkins, 1991.

Galison, Peter. *Einstein's Clocks, Poincaré's Maps: Empires of Time*. New York: W.W. Norton, 2003.

Gange, David. *Dialogues with the Dead: Egyptology in British Culture and Religion, 1822–1922*. Oxford: Oxford University Press, 2013.

Garwood, Christine. *Flat Earth: The History of an Infamous Idea*. New York: St Martin's Press, 2007.

Gawrych, George. *The Young Atatürk: From Ottoman Soldier to Statesman of Turkey*. London: I.B. Tauris, 2013.

Gillard, Derek. "Education in England: A Brief History," at http://www.educationengland.org.uk/history/chapter03.html (accessed 11 October 2019).

Glennie, Paul, and Nigel Thrift. *Shaping the Day: A History of Timekeeping in England and Wales, 1300–1800*. Oxford: Oxford University Press, 2009.

Green, Constance. *Washington: Village and Capital, 1800–1878*. Princeton, NJ: Princeton University Press, 1962.

Green, Lorne. *Chief Engineer: Life of a Nation Builder – Sandford Fleming.* Toronto: Dundurn Press, 1993.

"Greenwich Time All Over the World." *Leisure Hour* 34 (1885): 66.

Grier, David Alan. *When Computers Were Human.* Princeton, NJ: Princeton University Press, 2005.

Hacking, Ian. *The Taming of Chance.* Cambridge: Cambridge University Press, 1990.

Hawking, Stephen. *A Brief History of Time.* New York: Bantam Books, 1998.

Headrick, Daniel. *When Information Came of Age: Technologies of Knowledge in the Age of Reason and Revolution, 1700–1850.* Oxford: Oxford University Press, 2002.

Heathorn, Stephen. *For Home, Country, and Race: Constructing Gender, Class, and Englishness in the Elementary School, 1880–1914.* Toronto: University of Toronto Press, 2000.

Higgitt, Rebekah. "A British National Observatory: The Building of the New Physical Observatory at Greenwich, 1889–1898." *British Journal for the History of Science* 47, no. 4 (2014): 609–35.

Howse, Derek. *Greenwich Time and the Discovery of the Longitude.* Oxford: Oxford University Press, 1980.

International Meridian Conference (IMC), Washington, DC, 1884. See *Protocols of the Proceedings.*

Irokawa, Daikichi. *The Culture of the Meiji Period.* Princeton, NJ: Princeton University Press, 1985.

Jalland, Pat. *Women, Marriage, and Politics, 1860–1914.* Oxford: Oxford University Press, 1987.

Johnston, Scott. "The Construction of Modern Timekeeping in the Anglo-American World, 1876–1913." PhD dissertation, Department of History, McMaster University, 2018, at https://macsphere.mcmaster.ca/bitstream/11375/23324/2/Johnston_Scott_A_2018August_PhD.pdf (accessed 26 April 2021).

– "Managing the Observatory: Discipline, Order, and Disorder at Greenwich, 1835–1933." *British Journal for the History of Science* 54, no. 2 (2021): 155–75.

Keim, Randolph. *Society in Washington: Its Noted Men, Accomplished Women, Established Customs, and Notable Events.* Washington, DC: Harrisburg, 1887.

Kern, Stephen. *The Culture of Time and Space, 1880–1918.* London: Harvard University Press, 1983.

Knorr-Cetina, Karin. *The Manufacture of Knowledge: An Essay on the Constructivist and Contextual Nature of Science.* Oxford: Pergamon Press, 1981.

Landes, David. *Revolutions in Time: Clocks and the Making of the Modern World*. London: Harvard University Press, 1983.

Lankford, John. "Amateurs versus Professionals: The Controversy over Telescopes Size in Late Victorian Science." *Isis* 72, no. 1 (March 1981): 11–28.

Latour, Bruno, and Steve Woolgar. *Laboratory Life: The Construction of Scientific Facts*. Princeton, NJ: Princeton University Press, 1987.

Laughton, J.K. "Snow, William Parker." *Oxford Dictionary of National Biography* (2004), at https://doi.org/10.1093/ref:odnb/25980 (accessed 27 March 2019).

Launay, Françoise. *The Astronomer Jules Janssen: A Globetrotter of Celestial Physics*, trans. Storm Dunlop. New York: Springer, 2012.

Levine, Philippa. *The Amateur and the Professional: Antiquarians, Historians and Archaeologists in Victorian England, 1838–1886*. Cambridge: Cambridge University Press, 1986.

Levine-Clark, Marjorie. *Beyond the Reproductive Body: The Politics of Women's Health and Work in Early Victorian England*. Columbus: Ohio State University Press, 2004.

Lorimer, Douglas. *Science, Race Relations and Resistance: Britain, 1870–1914*. Manchester: Manchester University Press, 2013.

Lykknes, Annette, Donald Opitz, and Brigitte Van Tiggelen, eds. *For Better or Worse: Collaborative Couples in the Sciences*, London: Springer, 2012.

Mackenzie, Donald. *Statistics in Britain, 1865–1930: The Social Construction of Scientific Knowledge*. Edinburgh: Edinburgh University Press, 1981.

Mackenzie, John. *Propaganda and Empire: The Manipulation of British Public Opinion, 1880–1960*. Manchester: Manchester University Press, 1986.

Mansfield, Peter. *The British in Egypt*. New York: Holt, Rinehart and Winston, 1972.

Martin's Tables, or, One Language in Commerce. London: T. Fisher, Unwin, 1906.

Maunder, Annie, and Walter Maunder. *The Heavens and Their Story*. London: Robert Culley, 1908.

Maunder, Walter [and Annie Maunder, uncredited]. *The Astronomy of the Bible: An Elementary Commentary on the Astronomical References of Holy Scripture*. Bungay: Richard Clay and Sons, 1908.

Meadows, A.J. "Lockyer, Sir Joseph Norman, 1836–1920." *Oxford Dictionary of National Biography* (2006), at https://doi.org/10.1093/ref:odnb/34581 (accessed 12 May 2020).

Miller, J.R. *Shingwauk's Vision: A History of Native Residential Schools*. Toronto: University of Toronto Press, 1996.

Mitchell, Timothy. *Rule of Experts: Egypt, Techno-Politics, Modernity.* Berkeley: University of California Press, 2002.

Nanni, Giordano. *The Colonisation of Time: Ritual, Routine, and Resistance in the British Empire.* Manchester, Manchester University Press, 2012.

Ogilvie, Marilyn. "Obligatory Amateurs: Annie Maunder (1868–1947) and British Women Astronomers at the Dawn of Professional Astronomy." *British Journal for the History of Science* 33 (2007): 67–84.

Ogle, Vanessa. *The Global Transformation of Time, 1870–1950.* London and Cambridge: Harvard University Press, 2015.

O'Malley, Michael. *Keeping Watch: A History of American Time.* Washington, DC: Smithsonian Institute Press, 1990.

Oreopoulos, Philip. "Canadian Compulsory School Laws and Their Impact on Educational Attainment and Future Earnings." Family and Labour Studies Division, Statistics Canada and Department of Economics, University of Toronto. Catalogue No. 11F0019MIE, No. 251 (2005): 1–41.

Parramore, Lynn. *Reading the Sphinx: Ancient Egypt in Nineteenth Century Literary Culture.* New York: Palgrave MacMillan, 2008.

Perry, John. *The Story of Standards.* New York: Funk and Wagnalls, 1955.

Petrie, W.M. Flinders. *The Pyramids and Temples of Gizeh.* London: Field and Tuer, 1883.

Pietsch, Tamson. *Empire of Scholars.* Manchester: Manchester University Press, 2013.

Pionke, Albert. *The Ritual Culture of Victorian Professionals: Competing for Ceremonial Status, 1838–1877.* London: Routledge, 2013.

Porter, Bernard. *The Absent-Minded Imperialists: Empire, Society, and Culture in Britain.* Oxford: Oxford University Press, 2006.

Porter, Theodore. *Trust in Numbers: The Pursuit of Objectivity in Science and Public Life.* Princeton, NJ: Princeton University Press, 1995.

Potter, Simon. "Webs, Networks, and Systems: Globalization and the Mass Media in the Nineteenth- and Twentieth-Century British Empire." *Journal of British Studies* 46, no. 1 (2007): 621–46.

Prentice, Alison, and Susan Houston, eds. *Family, School, and Society in Nineteenth-Century Canada.* Toronto: Oxford University Press, 1975.

Protocols of the Proceedings: International Conference Held at Washington for the Purpose of Fixing a Prime Meridian and a Universal Day, October 1884. Washington, DC: Gibson Bros., 1884, at https://www.gutenberg.org/files/17759/17759-h/17759-h.htm (accessed 24 April 2021).

Pycior, Helena, Nancy Slack, and Pnina Abir-Am, eds. *Creative Couples in the Sciences.* New Brunswick, NJ: Rutgers University Press, 1996.

Ratcliff, Jessica. *The Transit of Venus Enterprise in Victorian Britain.* New York: Routledge, 2016.

Reisenauer, Eric. "'The Battle of the Standards': Great Pyramid Metrology and
    British Identity, 1859–1890." *Historian* 65, no. 4 (2003): 931–78.
Rooney, David. *Ruth Belville: The Greenwich Time Lady.* London: National
    Maritime Museum, 2008.
Rooney, David, and James Nye. "'Greenwich Observatory Time for the Public
    Benefit': Standard Time and Victorian Networks of Regulation." *British
    Journal for the History of Science*, 42, no. 1 (March 2009): 5–30.
Rudy, Jarrett. "Do You Have the Time? Modernity, Democracy, and the
    Beginnings of Daylight Saving Time in Montreal, 1907–1928." *Canadian
    Historical Review* 93, no. 4 (2012): 531–54.
"Rules and Regulations of the Mississaugas of the Credit." 1884.
Said, Edward. *Orientalism.* London: Routledge, 1978.
Sanders, L.C. "Beckett, Edmund [formerly Edmund Beckett Denison], first
    Baron Grimthorpe." *Oxford Dictionary of National Biography* (2004), at
    https://doi.org/10.1093/ref:odnb/30665 (accessed 11 July 2020).
Schaffer, Simon. "Astronomers Mark Time: Discipline and the Personal
    Equation." *Science in Context* 2, no. 1 (1988): 115–45.
Shapin, Steven. "Placing the View from Nowhere: Historical and Sociological
    Problems in the Location of Science." *Transactions of the Institute of British
    Geographers* 23, no. 1 (1998): 5–12.
Shively, Donald. *Tradition and Modernization in Japanese Culture.* Princeton,
    NJ: Princeton University Press, 1971.
Smith, Donald. *Sacred Feathers: The Reverend Peter Jones (Kahkewaquonaby)
    and the Mississauga Indians.* Toronto: University of Toronto Press, 1987.
Smyth, Charles Piazzi. *Our Inheritance in the Great Pyramid.* London: Isbister
    & Co., 1874.
Snow, William Parker. "Ocean Relief Depots." *Chambers's Journal of Popular
    Literature, Science, and Art*, no. 833 (27 Nov. 1880): 753–5.
"Some Foreign Ministers at Washington: Eight Portraits." *Phrenological
    Journal and Science of Health* 84 (Aug. 1887): 63–5.
Stephens, Carlene. "Before Standard Time: Distributing Time in 19th Century
    America." *Vistas in Astronomy* 28 (1985): 113–18.
Stewart, R.M. "The Time Service at the Dominion Observatory." *Journal of the
    Royal Astronomical Society of Canada* 1 (1907): 85–103.
Stone, G., and T.G. Otte. *Anglo-French Relations since the Late Eighteenth
    Century.* London: Routledge Taylor & Francis Group, 2009.
Tabili, Laura. "A Homogeneous Society? Britain's Internal 'Others,' 1800–
    Present." In *At Home with the Empire: Metropolitan Culture and the
    Imperial World*, ed. Catherine Hall and Sonya Rose, 53–76. Cambridge:
    Cambridge University Press, 2006.

– *We Ask for British Justice: Workers and Racial Difference in Late Imperial Britain*. Ithaca, NY: Cornell University Press, 1994.

Thomas, Morley. *The Beginnings of Canadian Meteorology*. Toronto: ECW Press, 1991.

Thompson, Andrew. *The Empire Strikes Back? The Impact of Imperialism on Britain from the Mid-Nineteenth Century*. Harlow: Pearson Education Limited, 2005.

Thompson, E.P. "Time, Work-Discipline, and Industrial Capitalism." *Past and Present* 38 (Dec. 1967): 56–97.

Thomson, Malcolm. *The Beginning of the Long Dash: A History of Timekeeping in Canada*. Toronto: University of Toronto Press, 1978.

Thoron, Ward, ed. *First of Hearts: Selected Letters of Mrs. Henry Adams 1865–1883*. San Francisco: Willowbank Books, 2011.

Tomalin, Marcus. *The French Language and British Literature, 1756–1830*. New York: Routledge, 2016.

Tosh, John. *Manliness and Masculinities in Nineteenth-Century Britain*. Harlow: Pearson Education, 2005.

Valrose, Viscount (pseud.). *Hon. Uncle Sam*. New York: John Delay, 1888.

Vetch, R.H. "Donnelly, Sir John Fretcheville Dykes." *Oxford Dictionary of National Biography* (2004), at https://doi.org/10.1093/ref:odnb/32861 (accessed 8 April 2021).

Vickery, Amanda. "Golden Age to Separate Spheres? A Review of the Categories and Chronology of English Women's History." *Historical Journal* 36, no. 2 (1993): 383–414.

Videira, Antonio. "Luiz Cruls e o Premio Valz de Astronomia." *Chronos* 7, no. 1 (2014): 85–104.

Visram, Rozina. *Asians in Britain: 400 Years of History*. London: Pluto, 2002.

Watts, Ruth. *Women in Science: A Social and Cultural History*. London: Routledge, 2007.

Weaver, John. *The Great Land Rush and the Making of the Modern World, 1650–1900*. Montreal: McGill-Queen's University Press, 2006.

Willinski, John. *Learning to Divide the World: Education at Empire's End*. Minneapolis: University of Minnesota Press, 1998.

Wilson, Margaret, *Ninth Astronomer Royal: The Life of Frank Watson Dyson*. Cambridge: W. Heffer & Sons, 1951.

Wise, Norton, ed. *The Values of Precision*. Princeton, NJ: Princeton University Press, 1997.

Wishnitzer, Avner. *Reading Clocks, Alla Turca: Time and Society in the Late Ottoman Empire*. Chicago: University of Chicago Press, 2015.

Withers, Charles. "Scale and the Geographies of Civic Science: Practice and
    Experience in the Meetings of the British Association for the Advancement
    of Science in Britain and in Ireland, c. 1845–1900." In *Geographies of
    Nineteenth-Century Science*, ed. David Livingstone and Charles Withers.
    Chicago: University of Chicago Press, 2011, 99–122.
– *Zero Degrees: Geographies of the Prime Meridian*. London and Cambridge:
    Harvard University Press, 2017.
Zeller, Suzanne. *Inventing Canada: Early Victorian Science and the Idea of a
    Transcontinental Nation*. Toronto: University of Toronto Press, 1987.

Index